Dreaming Ahead of Time

ALSO BY GARY LACHMAN

*Aleister Crowley: Magick, Rock and Roll,
and the Wickedest Man in the World*

Beyond the Robot: The Life and Work of Colin Wilson

The Caretakers of the Cosmos: Living Responsibly in an Unfinished World

A Dark Muse: A History of the Occult

Dark Star Rising: Magick and Power in the Age of Trump

The Dedalus Book of the 1960s: Turn Off Your Mind

The Dedalus Book of Literary Suicides: Dead Letters

The Dedalus Occult Reader: The Garden of Hermetic Dreams

Introducing Swedenborg: Correspondences

Jung The Mystic: The Esoteric Dimensions of Carl Jung's Life and Teachings

Lost Knowledge of the Imagination

Madame Blavatsky: The Mother of Modern Spirituality

Politics and the Occult: The Right, the Left, and the Radically Unseen

*The Quest for Hermes Trismegistus: From Ancient Egypt
to the Modern World*

*The Return of Holy Russia: Apocalyptic History, Mystical Awakening,
and the Struggle for the Soul of the World*

*Revolutionaries of the Soul: Reflections on Magicians,
Philosophers, and Occultists*

Rudolf Steiner: An Introduction to his Life and Work

In Search of P. D. Ouspensky: The Genius in the Shadow of Gurdjieff

A Secret History of Consciousness

The Secret Teachers of the Western World

Swedenborg: An Introduction to His Life and Ideas

*Turn Off Your Mind: The Mystic Sixties
and the Dark Side of the Age of Aquarius*

WRITTEN AS GARY VALENTINE

*New York Rocker: My Life in the Blank Generation
with Blondie, Iggy Pop and Others, 1974–1981*

Dreaming
Ahead
of Time

Experiences with Precognitive Dreams, Synchronicity and Coincidence

GARY LACHMAN

Floris
Books

To time-haunted men and women everywhere... and when.

'Lord, help me believe in the primary dreams of which my life is made.'
(Heard while in the hypnagogic state.)

First published by Floris Books in 2022
© 2022 Gary Lachman

Gary Lachman has asserted his right under the
Copyright, Design and Patents Act 1988
to be identified as the Author of this Work

 Also available as an eBook

British Library CIP Data available
ISBN 978-178250-786-4
Printed and bound in Great Britain by CPI Group (UK) Ltd, Croydon CR0 4YY

 Floris Books supports sustainable forest management by
printing this book on materials made from wood that
comes from responsible sources and reclaimed material

MIX
Paper from
responsible sources
FSC
www.fsc.org   FSC® C171272

Contents

Introduction
Of Hypnagogia and Hedgehogs

A talk at Brompton Cemetery. Liminal sleep. Prediction, prophecy,
premonition, precognition. I dream the future – and so do you.
Dunne, Priestley, Lethbridge.
A tweet about time.

The idea for writing this book came from a talk I gave in Brompton Cemetery in London in the spring of 2019. The talk, given as part of a series of events entitled 'On the Borderlands of Sleep', was on 'Hypnagogia and Precognitive Dreams'.[1] I have written about hypnagogia and 'hypnagogic hallucinations' in some of my other books; readers familiar with those will, I trust, pardon some repetition here.[2]

The hypnagogic state is the curious liminal condition each of us enters at least twice a day, when falling asleep and when waking up. There is some debate about the differences between these two transitional states, but this really shouldn't concern us.[3] The hypnagogic state is a 'half-dream' state, as the esoteric philosopher P. D. Ouspensky called it; it is midway between our usual waking state and unconsciousness. Most of us pass through this midway point without recognising it, but with practice we can learn how to remain in it, hovering, as it were, between full wakefulness and sleep. While on this threshold we can have some unusual experiences. We can watch dreams form, and see them with a startling vividness, while yet still being awake and aware of our surroundings. We are in two states of consciousness at the same time, simultaneously conscious and unconscious – or at least conscious enough to be critically aware of the symbols and images we associate with dreams, and that we believe have their source in the unconscious mind.

One of the characteristics of this unusual state is that it is 'auto-symbolic', as the early Freudian Herbert Silberer discovered.

That is, the images, symbols, and voices one hears in this condition are, or can be, symbolic of one's physical, emotional, and mental state at the time. Let me point out that I am aware of the paradox of being 'consciously unconscious', and of the difficulties involved in talking about 'the unconscious' as if it were a concrete thing, a part of us in the same way that our arms or legs are. But I should warn readers that if paradoxes put you off, you may well want to find another book. As we go along, we will run into more than a few of them here.

'Hypnagogia' is the name coined by the psychologist and philosopher Andreas Mavromatis, who wrote an exhaustive book on the subject, for the visions, images, symbols and auditory hallucinations that accompany this curious state. It is also a state linked to a variety of paranormal or 'psychic' phenomena, one of which is precognition.[4] Hypnagogic visions are also related to various esoteric ideas about consciousness; not a few of the figures in the history of esotericism and the exploration of inner states that I have written about – such as C. G. Jung and Swedenborg – were well practised hypnagogists.[5]

Precognition

The simplest definition of precognition is that it is having knowledge of some event before it happens. It is having knowledge about something before you should have it, before it was, by all ordinary standards, even *possible* for you to have it. If the definition of cognition is 'the mental process involved in knowing, learning, and understanding things' – from a dictionary taken at random – and if the prefix 'pre' means 'before' or 'previous to', then 'precognition' is the act of knowing something before we know it. And this is something that science, as well as reason and logic, tells us is impossible.

Sleep, or being on your way to it, is not a precondition of precognition. There are many examples of premonitions or presentiments coming to people while they were wide awake. But sleep and the half-sleep states of hypnagogia, do seem partial to precognition. I should also point out that there is a difference between precognition and premonitions or presentiments, as well as between precognition and prediction and prophecy, although of course they are all related.

In predictions and prophecies, someone makes a public claim about what will happen in the future, and this can be corroborated or dismissed when that future arrives. Prophecy is usually related in some way to the supernatural and religion. We hear about the 'fulfilment of prophecy' when an event seems to coincide with what the prophets of whatever religion had 'foretold'. Or there are the prophecies of visionaries like the sixteenth century physician and astrologer Nostradamus, whose *Centuries* seemingly foresaw the French Revolution and the rise of Hitler, among other things. In the twentieth century, Edgar Cayce, the American 'sleeping prophet', made similar pronouncements. Predictions tend to be secular – financial wizards often predict where the market is heading – and every year celebrity astrologers take a shot at foreseeing what is in store for the next twelve months. Premonitions and presentiments have more to do with a feeling that *something* is 'going to happen', usually something bad. These range from 'hunches' about some decision, to someone having a 'bad feeling' about some enterprise, to foreboding and full panic about a felt upcoming disaster.

Precognition is involved in premonitions and presentiments, but in the form of it that I will be paying most attention to, the person experiencing precognition *doesn't know* he or she is having a precognitive experience. Until, that is, the event they unconsciously pre-cognised comes to pass and they recognise that they already knew it would happen. There are hundreds of accounts of people waking from a dream and *knowing* that something has happened to a friend or loved one, or is about to. They see them in an accident or involved in some disaster or in some other form of danger. They may shrug the dream aside, or they may be moved by it enough to call the person. Sometimes their warning comes in time and the danger is averted. Other times, they are too late, or the friend ignores the warning, and the disaster comes to pass.

The kind of precognition I will be focusing on isn't this sort. The people experiencing it do not know until after the fact that they have glimpsed the future. Their glimpse of what is in store is most often mixed in with random thoughts, memories, vague reflections and other assorted confused matter, and does not, as in more dramatic premonitions, stand out as something significant. It is only when the future of which they have had a peek turns up that they say 'I knew this would happen! I saw it!' And this happens most often in dreams. Why that is so is still unclear, but more than one study suggests that the majority of precognitive experiences happen in dreams or, as Mavromatis recorded, in hypnagogic, 'half-dream' states.

My dream journals

I can agree with this because, with a few exceptions, my own precognitive experiences have happened in dreams. By now they have happened more times than I can remember, although I have been recording my dreams, off and on, for the past forty years.

The dreams included here come from that record. In fact, while going over my dream journals from the early 1990s in preparation for this book, I came across dreams that at the time did not seem precognitive but which now, many years later, seem likely candidates. The earliest precognitive dreams I recorded go back to 1980; the most recent are from the past few years. I may have had precognitive dreams earlier than 1980. I know I certainly dreamed a lot before then. If so they are lost in time – if, indeed, anything is truly lost in it.

I am not of a sensational nature. I try as a writer and a person to keep an even keel, to stay as balanced and composed as possible, in the face of often surprising events, and to assess things with a level head. But I can't deny it. I have dreamed the future. And if some of the theories about dreams and precognition we will look at are accurate, so have you.

Why now?

But back to my talk. As mentioned, I had written about hypnagogia in some of my books, and also in an article for the *Fortean Times* many years ago.[6] I had also written about a few of my precognitive dreams and related phenomena, such as 'synchronicities', the name the psychologist C. G. Jung gave to the odd experience of 'meaningful coincidence', in an article for *The Quest* magazine back in 1997.[7] Over the years, here and there I mentioned my precognitive dreams and synchronicities, but aside from that article, I did not write about them specifically. But in the back of my mind was the nagging knowledge that at some point I would have to.

Why would I have to write about them? For one thing, in several of my books I write about paranormal phenomena and, given that my 'future dreams' and synchronicities are the most convincing and frequent experience I have had of anything paranormal, it seemed natural that I should include them in my accounts, if only as examples of 'first hand' evidence.

I am aware that my account remains solely 'anecdotal', and hence inadmissible as scientific evidence for precognition or anything else. I can only say that I agree with many writers on the subject that anecdotal material is invariably more convincing than the most certified and approved scientific evidence – notwithstanding that, according to recent reports, there is quite a lot of this evidence around. Precognition, synchronicity, and odd coincidences have more to do with our everyday experience than with the artificial conditions of an experimental lab. Yet even there, in conditions often guaranteed to inhibit it, the paranormal – or whatever we'd like to call those strange bits of experience that do not fit into the usual pattern – make surprising appearances.

But the main reason why I would eventually have to write about these things at greater length than I have is their sheer mystery. How can precognition happen? How can we have knowledge of an event that has not yet happened? Or, if, in some way we are not yet able to explain, it *has* already happened, what does this say about our sense of having free will? Are our lives predestined, laid out before us, as railways are for the trains that run on them? If so, then precognition would seem more a curse than a blessing. Knowing in advance what will happen can lead to a depressing fatalism, the belief that our actions and decisions are not ours, but merely the acting out of an inescapable script, written by some unknown author. In 'Funes the Memorious', Jorge Luis Borges, a writer more than a little obsessed with time, shows the fate of someone who knows everything, all that has happened, is happening, and will happen.[8] It's not an enviable awareness: Funes' knowledge leads to paralysis and utter passivity before what is unavoidable. Yet, in more than one case, precognition has had the opposite effect, enabling people to avoid fate, even to change it, to act creatively, rather than as a mere puppet, moved by powers over which they had no control.

If only one case of precognition is accepted as true, then our usual ideas about time are surely in need of a review. My experiences with precognition suggest that this is so. Our ideas about time are in need of revision.

Time-haunted men

This was something that I tried to put across to my audience at my talk, held in a chapel in the cemetery on a chilly spring afternoon. After taking them through various accounts of hypnagogia from philosophers such as Aristotle and Jean-Paul Sartre, to psychologists such as Wilson Van Dusen and Julian Jaynes, to poets such as Coleridge and Blake, as well as some experiences of my own, I mentioned the paranormal phenomena associated with this strange liminal state. Telepathy, clairvoyance, shared or mutual dreams and other unusual mental experiences occur very frequently in hypnagogic states, as does perhaps the strangest paranormal experience of all, precognition. To end the talk I mentioned a few precognitive dreams of my own and discussed the ideas of three explorers of this mysterious phenomenon, whose work has informed much of this book: the writer J. B. Priestley, who called himself a 'time-haunted man;' T. C. Lethbridge, a Cambridge archaeologist and paranormal investigator, best known for his work with pendulums; and the most famous name associated with precognitive dreams, the aeronautics engineer and time theorist, J. W. Dunne.

It was in fact through reading Dunne's once famous book, *An Experiment With Time*, first published in 1927, that I began my own exploration into precognition. In closing the talk I mentioned Dunne's simple method to verify his belief that we all 'dream the future'. All that is necessary is that we record our dreams as soon as we remember them, and that we pay attention to what happens in our waking lives – good advice at any time. If we are serious about this we will discover, as Dunne did, that impossible or not, we do dream the future. It may not be in the exact form the future will take, although often enough it is, and may appear subject to a kind of 'symbolic distortion', which is not surprising as dreams speak a pictorial language made of symbols, images, metaphors and, very often, puns and plays on words. But the symbolism never veers very far from the fact – or future fact. I suggested that anyone in the audience who would like to verify this could do so that night. All they had to do was to write down their dreams the next morning, and see what happened.

A tweet in time

At least one person at the talk took me at my word. The next day, logging on to Twitter, I saw that someone in the audience had tweeted about the talk. She couldn't believe it, she said, but it was true. She *had* dreamed the future. She wrote that she had dreamed that she had picked a hedgehog up out of the road, where it could have been run over, and put it on the pavement, where it was safe. That morning, logging on to her Twitter feed, the first thing she saw was a post about 'keeping hedgehogs safe', about protecting them from traffic and other urban dangers. 'OMG,' she wrote, 'it's true!' I replied, writing that 'This is the kind of symbolic distortion that happens with precog dreams. You didn't actually save a hedgehog, but you saw a post about them being endangered and how you can help.' I added, 'I'm sure they are appreciative.'

I don't know if this person continued writing down her dreams, I never heard from her again, but I hope she has. But this tweet alone – assuming it was not a prank – was enough to convince me that I needed to look back over my dreams and finally make an attempt to understand exactly what was happening in them. This book is the result. And I'll start by sharing some of my dreams with you.

Chapter One
I Had a Dream Last Night

Dream and reality. King Kong. The Occult *on the Bowery. Faculty X.*
'Presence, Dear'. An Experiment With Time. *Cosmic consciousness.*
The Shadow. *Coppola's* Dracula. *Dreams and film. Synchronicity*
clouds. Triviality of most precognitive dreams.
A premonitory alertness.

The dreams, synchronicities, and odd coincidences discussed in this chapter come from journals I have been keeping since the 1980s. Some periods are better represented than others, and at different times my dream life seems simply to have stopped, or at least I have gone through long patches, sometimes months, when, try as I may, I do not remember my dreams. This is often compensated for by stretches when my dreams are so long and complicated, so odd and unusual, that it takes some time to record them. Some of the dreams here come from a period like this, when my dream life was so complex and involved that I had to stop paying attention to it, simply so that I'd have time in the morning for something other than recording my dreams. These dreams involved strange adventures, magical battles – more than once I have fended off Satanists and witches – visits from aliens, incredible journeys, and 'meetings with remarkable men' that would leave me half exhausted and utterly convinced that in my dreams I had actually taken part in these events. I had the feeling that these dreams took place not solely 'in my head', but were in some way 'real' in more than a psychological sense. I felt that in them I had visited some interior terrain different from the more common 'everynight' dreams that have me worried about some trivial problem, undressed in public, or anxiously missing a train. If dreams serve a compensatory function, as C. G. Jung, one of the great dreamers and dream interpreters of the last century believed they did, evening out imbalances in our waking mind, then I can only assume that at this time of my life, I must have been rather bored and frustrated. And having read some of the journal entries that accompanied the records of my dreams at this time – leading up to when I left Los Angeles at the end of 1995 and moved to London to start a new life as a full time writer – I have to say it appears I was.

That I still have dreams whether I remember them or not – that we all do – seems to have been confirmed by science. According to several studies, we dream throughout the night; although many people claim that they don't dream, this is untrue. They may not remember their dreams – not everyone has the same power of dream recall, although it can be developed with practice – but they have them nevertheless. Dreaming is something we do and we continue to do it throughout the sleeping hours of our life, in the same way that we continue to breathe and our hearts continue to beat. And if some investigators into the dark side of the mind are correct, we also dream throughout the day. Along with the daydreams with which we are familiar, it seems that those we associate with the night are still at work as we go about our daily business.

With this in mind, the idea that our whole life is a dream, that the world we wake up to from sleep is as much a dream as the ones we have left behind, and which informs spiritual teachings and practices such as Tibetan Buddhism and the Fourth Way of Gurdjieff and Ouspensky, takes on a new and significant meaning.

King Kong and UFOs

I know I dreamed a great deal as a child, but unlike some dreamers – again, Jung is a good example – I cannot today recall dreams from my childhood. The earliest dreams I can recall are from my late teens, when I was first living in New York in the mid-1970s. Two dreams I still remember vividly from that time. In one, I was sitting on King Kong's shoulder, clutching tightly to his fur as he swam around Manhattan. I was nineteen and just starting a career as a musician, in the early days of what would be called 'punk rock'. I can recall the tremendous sense of power and exhilaration as Kong cut through the Hudson, and the fantastic skyline of a dream New York, with skyscrapers of shining chrome and glass, taller than the real ones. In another dream, huge flying saucers, massive UFO 'mother ships', slowly drift over New York, preparing to make contact, as we have since seen in dozens of films, although *Close Encounters of the Third Kind* had yet to be made. I watched them glide over the tops of the buildings, their shadow covering city blocks. Again, there was a terrific sense of exhilaration and wonder, the feeling that 'They're here! They're really here!' and astonishment that the first contact with extra-terrestrials was about to be made. (As you might suspect, when I awoke, I was terribly disappointed that this hadn't happened.)

What stood out from these dreams, aside from their fantastic content, was their remarkable clarity and vividness. King Kong's fur was as real as that of a dog I might pat. The undersides of the UFO were as intricately patterned as an electrical circuit. And this has been true of many dreams at different times in my life: they seemed to possess a reality *greater* than that of waking life. In his researches into dreams, Ouspensky remarked on the incredible talent of the 'dream artist', our inner stage director who is able to recreate reality so sharply that it is at times almost painful.[1] Today's efforts at virtual reality and HD TV pale by comparison and it is not surprising that many people, having experienced what I call the 'crackle of reality' in dreams, believe that what they have emerged from was more than 'only a dream'.

The Occult on the Bowery

Oddly enough, around the same time as I had these dreams, I first became interested in magic, the occult, mysticism and what I came to know as the 'esoteric tradition'. It was in the spring of 1975, when I was living on the Bowery in New York City with the singer and guitarist of a band I had recently joined.[2] The person renting us our floor was an artist of sorts, with an interest in the notorious dark magician Aleister Crowley. He used to give impromptu readings with the Crowley Thoth tarot deck, something of a rarity

at the time, and he painted large canvases based on the trump
cards; I even modelled for one of these. The singer and guitarist had
a kitschy interest in the occult. Along with pictures of the Ramones
and Velvet Underground album covers, voodoo dolls, inverted
pentagrams and crosses adorned the walls of our floor, as did several
Tibetan thangkas, Buddhist paintings used in meditation, one
of which depicted a deceased monk cheerfully being eaten by his
colleagues.

Until then I had no interest in magic or the occult, aside from
a love of weird fiction – think H. P. Lovecraft – and horror films
from the 1930s and 40s (one of which was indeed *King Kong*).
But in that atmosphere my interest was piqued and a book I came
across then literally changed my life. It was Colin Wilson's *The
Occult*. After reading it I became fascinated with magic, the occult
and the paranormal, and started what became a lifelong interest
in these things, reading everything I could get my hands on about
them. Some years later, this obsession turned into a new career as
a writer on these subjects. And eventually I wrote a book about
Wilson.[3]

Yet something else I had discovered in Wilson's book made
an even deeper impression on me. In it he talked about a strange
ability we all possess but are unaware that we do, a power that
enables us to step outside of time, to no longer be bound by the
limitations of the present, the prison of 'here and now', but to be
able to grasp the *reality* of 'other times and places', and to, in effect,
travel in time. He called this 'Faculty X', simply because we don't
have a name for it. Here I can just mention it, and we will return to
it later in this book.

Some friends found my new interest in the occult odd, but
I can't say it was unique. The 1970s saw an 'occult boom' and
publishers were putting out books on the subject by the dozen.
Bookshops were crowded with discount tables full of inexpensive
reprints of classical works on magic and mysticism, as well as
dozens of books about UFOs, 'ancient aliens', ESP, 'out of body'
experiences, reincarnation, the Hermetic Order of the Golden
Dawn, astral travelling, dreams, precognition, altered states of
consciousness and much that by now has become part of the 'new
age' and 'mind, body, spirit' culture of the twenty-first century.

I read all of it; I even got books out of the library. And soon after we started going out, my girlfriend, whom I'll call L., did too. One result of this was a song I wrote for her and which the band later recorded. '(I'm Always Touched by Your) Presence, Dear', became a UK and European hit in 1978 – I received a gold record for it – the one time, I think, that a song about telepathy and with the word 'theosophy' in its lyrics made the Top Ten.[4]

'Presence, Dear'

What had happened was that while I was on tour, L. and I realised that we were having similar dreams and that we seemed to be thinking about each other at the same time, a not uncommon occurrence with couples. As most studies show, an emotional link is the most important ingredient in any telepathic communication. In fact, at the time we were having these experiences, similar ones were being investigated at the Maimonides Dream Laboratory in Brooklyn, established by Montague Ullman and Stanley Krippner in the 1960s. Calling each other was not as easy as it is today – mobile phones were some years away – yet it seemed that whenever one of us tried to call, the other had the same thing in mind. The thing happened so often that when I returned to New York I thought about it quite a bit.[5] One afternoon I picked up my guitar and started strumming. Soon a song took shape. 'Presence, Dear' was among the new material we debuted in the spring of 1977, when we toured the UK during the Queen's Silver Jubilee, around the same time that the Sex Pistols' 'God Save the Queen' was getting banned by the BBC.

I left the band after that tour and moved to Los Angeles to be with L., because she wanted to pursue her acting career in Hollywood. There I started my own band, the Know, the name inspired by my interest in Gnosticism, and continued my reading. By 1980 L. and I had drifted apart, and I had moved the band back to New York. And it was there that my adventures with dreams really began.

An experiment with time

One of the books I had come across in my reading was called *An Experiment with Time*. Its author, as mentioned, was a hard-headed, no nonsense, aerodynamics engineer and military man, named John William Dunne. I had read about Dunne's ideas about time and dreams in *Mysteries*, Wilson's sequel to *The Occult*, and I had made it a practice to follow up the leads that Wilson so generously laid out (readers who do this wind up with the equivalent of a liberal arts education). I saw a copy of *An Experiment with Time* in a second-hand bookshop, bought it and brought it back to my flat. I was fascinated by Dunne's account of how he dreamed the future. For this was the reason for his 'experiment with time': to determine conclusively that he had seen the future in his dreams.

Soon after I finished the book, I decided that I would follow Dunne's suggestion and start writing down my dreams. Results came almost immediately. I was stunned. Dunne was right. I *did* dream the future. Just as the woman who would come to my talk forty years later discovered in her dream about a hedgehog, I was astonished to find that, impossible as it seemed, there was the future, of all places, in my dreams.

The results of my own experiment, though convincing, were not spectacular, as some of Dunne's early 'future dreams' were. Dunne began his experiment with time when he woke up from a dream in which he had been arguing with someone about the time. He said it was 4:30 am. The other person disagreed. In the dream Dunne assumed that his pocket watch must have stopped. He reached for it to check, and found that it had. He then woke up.

Dunne was intrigued enough by the dream to get up and check his watch. It *had* stopped and at precisely 4:30. Dunne assumed he must have noticed this earlier, forgotten it, and also forgotten to wind his watch, and his dream was prompted by this. He wound the watch and went back to bed. When he woke up later, he assumed he would have to adjust his watch, as he had no idea of the real time when he wound it in the night. When he checked the watch against a clock, he discovered that it had the correct time. It *had* stopped at 4:30, just as it had in his dream.

That was odd. But it was only the beginning.

Dunne decided to see if he could repeat the experience. Lying in bed one morning, he drifted into 'one of those semi-dozes in which one is aware of one's situation' – that is, a hypnagogic state – and tried to 'see' his watch. A moment later he did. The time was two and a half minutes past eight. He then opened his eyes and grabbed his watch. The time was 8:02 and a half, just as in his half-dream vision. There was 'no way out', Dunne decided. He was forced to accept that he 'possessed some funny faculty of *seeing* – seeing through obstacles, across space, and round corners.'[6] But this was more than a kind of clairvoyance, the term then used for what we today call 'remote viewing', a less romantic description prompted by its study and use for military purposes.[7] What Dunne was about to discover was that he could not only see 'round corners', but also round time.

Another dream clinched this. Dunne dreamed that he was in Egypt, near Khartoum. Three men, dressed in ragged army uniforms, those of Boers from South Africa, approached him from the south. In the dream he asked them why they should travel on foot from the bottom of the continent to the top. They replied that this was exactly what they were doing. That morning Dunne saw a newspaper headline, 'The Cape to Cairo: Expedition at Khartoum'; the story was about the expedition he had dreamed about the night before. But the most thrilling dream was of the kind that are most often mentioned when discussing precognitive or premonitory dreams, that is, one involving a disaster. Dunne dreamed he was on an island, and cracks and fissures were opening up under his feet. Steam escaped from them and Dunne realised that the island was about to explode. He knew everyone on the island had to be evacuated, and in the dream he made frantic efforts to get the French authorities – for they were the ones in charge – to recognise the danger and to make arrangements for the evacuation. Throughout the dream Dunne knew that 4,000 lives would be lost if swift action wasn't taken. The authorities ignored him, and Dunne woke from the dream as he was insisting to some obdurate French official that 4,000 lives were at stake.

At the time of this dream, Dunne was camped at a military base in what was then the Orange Free State. Newspapers arrived intermittently, and it wasn't until some time after his dream that

Dunne read about the eruption of Mount Pelée on the island of Martinique, then under French control, on May 8, 1902.[8] It was the worst volcanic eruption of the twentieth century, rivalling the eruption of Krakatoa in 1883. The town of St Pierre was completely destroyed. Once again, Dunne had dreamed the future. But what he came to see was that he did not dream of the eruption or of the expedition themselves, but of his *reading about them* in the newspaper. It wasn't until some time later that Dunne realised that his figure of 4,000 casualties was incorrect. The real number in the headline was 40,000. In actual fact, neither of these figures was correct, and subsequent reports made clear that the number of casualties was closer to 30,000. Where did the figure of 4,000 come from? Dunne realised it came from him *reading the headline incorrectly in the dream*, just as one might misread a headline in 'real life'.

What this and Dunne's subsequent future dreams led him to see was that the future that he glimpsed in his dreams was precisely his own. It was his own future experience that had somehow appeared in his dreams. He did not hover above the future 'in general', overlooking a temporal landscape that did not necessarily concern him personally, but saw what would happen to him in particular *ahead of time*. It seemed that some of his own future experience was somehow knocked out place and had, as it were, jumped the queue. Dunne concluded that if he had read about the Cape to Cairo expedition or the Martinique disaster as he would ordinarily, and *then* had the dreams, there would be nothing unusual about them. And although we can take argument with Dunne's causal account of dreams (I saw this yesterday and so I dreamed about it last night), his point is clear.[9] Dunne was having the experience he would have had, but in his future dreams, cause and effect were reversed, a back to front sequence that, according to some elementary particle physicists, can happen at the sub-atomic level. The effect – the dream – came *before* the cause – reading the newspaper headline. For Dunne, as for Shakespeare's Hamlet, time, it seemed, was out of joint.

I dream the future

And so it was, it seemed, for me. The first future dream I recorded after taking Dunne's advice was of playing a red guitar. I didn't own one, nor did I know anyone who did. But later that afternoon, having recorded my dream that morning, through a chance meeting with a friend, I wound up at the apartment of someone I didn't know. While there this new acquaintance handed me a red guitar and said, 'Check this out.' I started strumming and only then did I recall the dream. The next future dream was a bit more exciting. I had dreamed of sitting cosily with an attractive woman I knew slightly and whom I would not have minded getting to know better. I chalked that one up to Freudian wish-fulfilment. But then, through another chance meeting, I found myself unexpectedly in a rather intimate setting with the woman in question; Freud or not, that wish was fulfilled. Other dreams followed and I recorded them. But at some point over the years that particular dream journal got lost in the shuffle.

The next future dream I recorded comes from late 1981, when I had just finished a tour with the last band I played with and had decided to retire from music. In a dream I was riding on the back of a motor scooter with someone I didn't know through an unfamiliar neighbourhood. The landscape was hilly, with palm trees and other vegetation not native to Manhattan, where I was still living. At one point in the dream we passed some large trucks with the words 'cosmic consciousness' painted on their sides. Then I woke up.

This seems an odd enough dream to begin with, but some months later, in the spring of 1982, I found myself on the back of a Vespa, riding with a new friend through the Silverlake district of Los Angeles, where I had just moved back to from New York. We passed the hills I had seen in the dream, also the palm trees. But instead of the trucks emblazoned with 'cosmic consciousness' on their sides there was something else. As we passed a street with the same name as my friend I pointed this out to him, and as I did I remembered my dream. Suddenly a powerful feeling of *déjà vu* came over me, a sensation as if walls of some sort had fallen, and that my awareness of things had expanded in all directions. This seemed a taste of the 'cosmic consciousness' that the trucks in my dream may have been 'delivering' – an example, I suggest, of 'symbolic distortion'. It was also around this time that I had a very powerful lucid dream – dreams in which we know we are dreaming and yet continue to dream – involving this friend, which induced a similar feeling. I don't remember the details, but in the dream I turned to my friend and said, 'This is a dream, isn't it?' and he turned to me, smiled and said, 'Yes, it is.' At that point, as on the Vespa, there was a sudden sense of expanded horizons, of a kind of space opening up around me.

I can't say that this sense of 'expanded consciousness' accompanied many of my dreams, although I have felt it a few times since then, in dreams and in waking life. (My most intense 'peak experiences' though, I believe have come in dreams.) But I can say that there is a peculiar *feeling* associated with my future dreams. J. B. Priestley, that 'time-haunted man', spoke of 'the peculiar tone and tang of these experiences and the strange thrill of recognition' that accompanies them.[10] I agree with Priestley. Often, not always, there is a kind of premonitory alertness, a tingle that precedes my becoming aware that what I am experiencing now, in waking life, is something that I dreamed of the night before. And the arrival of the full recognition that this is so, is invariably accompanied by a laugh of surprise and joy. No matter how many times it has happened – and it has happened many times – the *reality* of having seen a few steps ahead of the present hits me and I am stopped in my tracks.

Telepathy or precognition?

Although in this 'cosmic consciousness' dream the precognitive element involved a span of some months, the time lag involved in most of my precognitive dreams is usually a day or two. Sometimes it can be only hours and fairly frequently only a matter of minutes. That is, I will awake from a dream and shortly after waking up, something in the dream I had just exited will appear. This seems to happen most often if I wake up earlier than usual and return to sleep for a short while: usually I will have two or three dream vignettes at this time. So, one morning I awoke from a dream in which I saw images of a car driving backwards, that is, backing up a long distance. Soon after waking I looked out of the window and saw a van pull out of a parking space and back up the length of the block. In another dream from the same time, I picked up my ex-wife's beret, tried it on, realised it was hers, and put it back. When F., my ex-wife, woke up soon after I did, she asked me if I had seen her beret. On another occasion, I dreamed that our bedroom was filled with swarms of gnats. When F. woke after I did, she complained of a mosquito bite.

It's possible that these last two dreams could be accounted for by telepathy. F. could have been thinking of her beret in her sleep and I could have picked up her thoughts and they could have turned into my dream. Her discomfort from the mosquito bite could have been communicated to me in the night and so influenced my dream. I could also have heard the mosquito's whining, and that could have prompted the dream. But I am a terribly light sleeper and am generally awakened by the slightest noise. And a mosquito buzzing past my ear usually has me out of bed and turning on the lights, determined to hunt it down.

But I have to say that on more than one occasion it seemed that F. and I were sharing dreams, as L. and I back in my musician days did. In one dream I found myself walking down a corridor, followed by a camel. A few days later F. told me that the night before she had a dream in which I was walking with a camel. I hadn't mentioned my camel dream to her, so this was sheer coincidence, a synchronicity, or in some way her sleeping mind picked up on my own camel dream and repeated it. On another occasion, I woke from another dream, this time about New York. When F. woke up shortly after, she told me she dreamed about New York too, before I had mentioned my dream. And again, in a dream I was at a lecture on Jung. A girl sitting next to me handed out sweets, chocolate, and cake. When F. woke she told me that she had dreamed of eating many sweets, cakes and cookies.

More recently, after waking up early and drifting back into sleep, I dreamed of a glass filled with a golden-brown liquid of some kind. That morning I tried a herbal drink a friend had suggested. When I reached for the glass after stirring the concoction, I recognised that it was the same colour as the liquid in my dream. I do not usually drink herbal teas and had not intended to try this one until that morning.

One rather interesting 'quick return' precognitive dream happened while I was reading material for this book. Here is how I recorded it in my journal:

> I dream that I am looking at a book, one of my own, and am surprised to see that there are pictures in it. When I look more closely I see that they move, like images on an old B&W television set. After I record this in my dream journal, I open Havelock Ellis' *The World Of Dreams,* which I am re-reading. The first sentence I read is 'the commonest kind of dream is mainly a picture, but it is always a living and moving picture…' So in a dream I see moving pictures in a book, and in a book I read that pictures in a dream move.

Triviality of precognitive dreams

The reader might note that in these examples the incidents involved fairly trivial items, nothing of any real importance or significance, although this last example suggests that my dream artist was aware of my research and was responding in a clever and creative way; it is also one of quite a few dreams involving books or something I am reading. But it is true of the majority of my future dreams that they involve fairly trivial sorts of things. The only thing significant about these dreams is that they were precognitive, although some precognitive dreams have been important in my relationships and my psychological development, as have non-precognitive ones, as might be expected. I have never had a future dream from which I woke up *feeling* that I had had a vision of the future, nor have I had one that was a premonition or presentiment of some disaster.

I did on one occasion have a precognitive dream about a volcano, as Dunne did. I awoke from a dream in which I was with my ex-wife, who is half-Japanese, and we were in Japan with her uncle. He was helping us to escape from a flood of lava and urging us to hurry. I had never met her uncle nor have I ever been to Japan, and I did not hear about the eruption of Mount Unzen in the Nagasaki Prefecture in June 1991 until a week after the disaster and my dream.[11] It is possible that some form of telepathy might account for this dream too. My ex-wife may have been in some kind of 'rapport' with her uncle, and I may have picked up her intuition, but she didn't mention this to me nor was she aware of the volcano until we both heard about it. Oddly enough, another dream of disaster involved Japan. I dreamed of tremendous explosions and devastated landscapes; only later that day did I realise it was August 6, the anniversary of Hiroshima. It seems that my dreaming mind, however, knew.

The Shadow knows

Yet although most of my future dreams are of experiences I will have shortly after dreaming about them, the most convincing of my precognitive dreams involved a period of years. Again, let me quote from my journal. This comes from 1994:

> On 15 March 1990, I recorded in my dream journal that in a dream I saw a film based on the pulp magazine character the Shadow. I read the Shadow stories as a kid and at the time I was deeply interested in Jung and was seeing a Jungian analyst, so having the two combine seemed interesting but not surprising; the 'shadow' of course is a central part of Jung's psychology. I recorded that in the dream, in one sequence of the film [here I am quoting from my 1990 journal] 'I watch the Shadow emerge from a wall; by this I mean that he literally was a shadow, that is two dimensional, and then solidified into a three-dimensional body. It was as if he was drawn on the wall and then stepped out of the picture.' I also record that the film had something to do with a ball or spherical object, and that in another sequence I see the Shadow's gloved hand holding a revolver (or pistol of some kind) before a wall. I wrote in my journal: 'He is creeping up on someone, and all we see is the slow movement of the pistol and the hand, and the shadow this throws on the wall.'

In July of 1994, F. and I went to see *The Shadow*,
starring Alec Baldwin. On the way, I recalled the dream
and mentioned it to F. Before the film started, I reminded
F. of one scene from the dream, the one where the Shadow
is two-dimensional, then steps off the wall, and becomes
three-dimensional. Halfway through the film, there was a
scene just like this. Also, the plot of the film follows the
Shadow's attempts to prevent power-crazed madmen from
blowing up New York with a nuclear weapon that was in the
shape of a sphere, a 'beryllium sphere'. Also in the film is a
scene in which the Shadow is in a hall of mirrors, and we
see a full screen shot of his .45 slowly moving in front of a
mirror – the reflection of the gun being the shadow 'thrown
on the wall'.

There were a few other odd connections. On the way to the
film, we ran into a couple that we knew but of whom we were
not particularly fond. We used to refer to them as our 'shadow', in
Jungian speak. Where did we run into them? At a tea shop called
Chado.[12]

What to make of this? I am not quite sure, but I am inclined to
think that the attention I was paying my unconscious through the
Jungian work I was doing had something to do with it. This kind
of 'overspill' from my dream world into my everyday one happened
fairly often during this period and brought the phenomena of
synchronicity and 'meaningful coincidence' into a curious and
sometimes disturbing blend with what was happening in my
sleeping mind. There were times when I felt I was walking in a kind
of 'synchronicity cloud', in which parts of my dream and everyday
life were fusing into some state that was neither one nor the other,
with coincidences and precognitive flashes going off around me
for a week at a time, accompanied by the feeling that the partition
between my inner and outer world had become extremely porous.

The Shadow was not the only film involving precognitive dreams
and my ex-wife. Around this time F. worked for Sony Studios, and
one of the films she was working on was Francis Ford Coppola's
version of *Dracula*. It was all very hush-hush and F. wasn't allowed
to talk about the film so she didn't. One night I had a frightful

nightmare in which I rescued a woman from some gruesome fate in a strange, Hieronymus Bosch-like village. Wolves hung upside down on crosses, fires burned, throwing up blood-red shadows and thick black smoke, and foul-looking characters waded through arms, legs and other body parts, carrying axes and pikes. I wrote in my journal that it was 'a charnel house, like something out of one of Swedenborg's hells.' (Swedenborg's depictions of hell are on a par with those of Dante). 'It was,' I wrote, 'truly evil.'

When the film was completed, there was a special screening of it, to which we were invited. No sooner had the screening started than I saw the bloody scene from my dream. The film begins with a sequence showing the handiwork of Vlad the Impaler, the fifteenth century Romanian sadist upon whom Bram Stoker based his story about Count Dracula. And I felt then the odd sensation, as I did later when seeing *The Shadow*, that I was starting to call the 'time slip', when, as Dunne did, I recognised that parts of my 'timeline' were somehow being lifted out of their 'normal' sequence and shifted days, months, and even years ahead of schedule.[13]

Dreams and films

Other future dreams involved films. The two examples I present here are a bit complicated and make liberal use of 'symbolic distortion'. Again, from my journal:

> In a dream, F. and I are going to a film. At the ticket booth, I hand the man my credit card. He looks at it, then at me, and asks who I am. I tell him I'm the same person whose name is on the card. He asks for my ID. I hand him my driver's license. I then realize that the names are a bit different on the cards; one has my middle name but not the other.[14] He excuses himself and consults his colleagues. One of them approaches me and breaks out into a wild dance, shouting and gesticulating. Then some security guards arrive.
>
> That afternoon, F. and I did try to see a film. But we ran into trouble. When we got to the theatre, I saw that it

resembled the one in my dream. I thought to tempt fate and try my luck with my credit card, but when I got to the booth I saw that the film we wanted to see wasn't playing there. I had got the theatre wrong. So this theatre didn't take my credit card, just as the one in my dream hadn't.

Later that evening, I saw that *The Great Escape*, a favourite film of my youth, was on television. I thought I would watch a few minutes of it, just for fun. It had been on for a while, and by the time I started watching, the Allied prisoners had escaped and were being hunted by the Germans. The prisoners were dressed in civilian clothes and were carrying forged identity papers. Many of the scenes were about Gestapo agents checking people's ID.

In the dream I had trouble getting into the theatre, and in reality I did. In the dream I had to show the authorities my identification papers, and in a film I watched that evening, having identity papers was a key element.

Here is another future dream concerning a film:

In a dream I have borrowed a Jewish friend's guitar. On my way to return it, I play it but in an odd way: using only one finger to hold the strings down against the neck. When I get to my friend's home, I see that some sort of Jewish celebration is going on. Then, at some point, my friend tells me that he must spend some time in a jail or cell in the basement. He can only avoid it if someone takes his place, and I offer to do that. Then, after seeing the cell, I change my mind. We leave and enter a large banquet hall, where a concert is about to be held. The pieces of music to be performed are by Arnold Schoenberg: *Verklarte Nacht* and *Music for a Film Accompaniment*.

A few days later, F. and I watch a film, *Swing Kids*, about German teenagers being forced to join the Nazi Youth. In the film, one Jewish youth is a jazz enthusiast and a fan of Django Rhinehart. He is also a guitarist. He is attacked by some Nazi youth who break his fingers. The idea is that he would never play guitar again. But he learns to play with

only two fingers, as Rhinehart did. The whole atmosphere
of the film was anti-Jewish and in the dream my friend,
who is Jewish, has to spend time in a prison. The two pieces
of music to be performed at the banquet are by Arnold
Schoenberg, a Jewish composer. On the recording I have of
his *Music for a Film Accompaniment* there is also his memorial
to the Holocaust victims, *A Survivor From Warsaw*.

So in this dream and the previous one, the precognitive elements
are not presented unadulterated; they are 'symbolically distorted'.
The person at the ticket booth in 'real life' didn't ask for my ID,
but there was trouble at the cinema and we didn't get in. Later,
watching a film, having ID was a central theme. This is the same in
the next dream. I am returning a guitar that I play with only one
finger to a Jewish friend, who tells me, during a Jewish celebration,
that he must spend time in a cell. Then, at a concert, two pieces of
music by a Jewish composer are going to be performed. In the film
I saw after the dream, a Jewish guitarist has his fingers broken but
learns how to play using only two. The whole atmosphere of the
film is anti-Jewish, and Jews are being carted off to concentration
camps. And although it isn't on the program for the dream concert,
Schoenberg, the Jewish composer in the dream, did write a famous
piece in memory of the Holocaust victims. And a recording I
have of his *A Survivor From Warsaw* also has his *Music for Film
Accompaniment*, which was on the dream program.[15]
Some readers may find calling these dreams precognitive
stretching it a bit, but for me they had that tang of recognition J. B.
Priestley talked about.

More future dreams

Yet most of my precognitive dreams are fairly simple and, as
mentioned, not particularly interesting. Let me relate a few. One
morning I dreamed of tiny frogs coming out of a hole in the floor
of my room. That afternoon, picking up my son from school,
his teacher asked to speak with me. I noticed he was carrying a

jar filled with water and asked what it was. 'Frog spawn,' he said, 'collected by the class.'

I wake from a dream in which I am on a ship at sea. That morning, a friend calls to ask if I would like to go sailing that afternoon, something I had not done before. In a dream, Data, the android from *Star Trek: The Next Generation*, and I have captured a dangerous criminal and are questioning him. In the dream, all I can see of Data is his head. In the next episode of *Star Trek* that I watch, Data's head is removed and used to control a computer. In a dream I come across a newspaper with a two-page article on rock music. In it is a section about me leaving my old band, with a photograph of me. That day a friend calls to tell me that my picture is in the *Los Angeles Times* in an article like the one in my dream. (I have never been a newspaper reader and so wouldn't have known of this unless the friend told me.) I dream of demolishing a wall, and of watching someone emerge from the rubble. Later that day, I spoke with my parents and they told me that they had to take a wall out in their new house in order to look for termites. In a more recent dream someone has an old revolver and is checking to see if it is loaded. 'Is there anything left in that?' someone asks. That evening I watch an episode of *Inspector Montalbano*, 'The Goldfinch and the Cat', in which a mugger fires blanks from an old revolver.

I wake with a faint memory of seeing grey and white 'lines' or 'strings' crossing each other, like hairs. Walking here in North London, I stop to look at the Regent's Canal and find myself looking straight through a spider's web. I dream that I am sitting at a picnic table and that R., my sons' mother, comes by on her bicycle and asks if I will 'pay' for her. That afternoon I am at a café, sitting at an outdoor table. R, who I hadn't seen in some time, rides by, sees me, and comes to the table. When I ask if she'd like a coffee, I recognise the table from the dream. It has wooden slats, as the picnic table in my dream did, and it is outside. R doesn't ask me to pay for her, but I do. On another occasion I dreamed of a huge salad; it was overflowing, too much for the bowl it was in. That morning, before I mentioned the dream, R. told me about a 'giant salad' her housemates made the night before.

Waking precognition

I could go on and indeed it is difficult to stop giving examples, such as these two of precognitive experiences I had in a waking state. In one I am awake, lying on my couch, listening to the radio, and I feel a sudden urge to hear Beethoven's *Pastoral Symphony*. Before getting up to put the CD on, I wait a moment to see what's next on the radio. It is Beethoven's *Pastoral Symphony*. On another occasion, I am walking on Rosslyn Hill, in Hampstead, not far from where I live in London, and recall that the biologist Rupert Sheldrake – of 'morphogenetic fields' fame – lives in the neighbourhood. I wonder if I'll see him. A moment later I do see him, walking towards me up the hill.

But most of my experiences of precognition have come in dreams, of which I cannot resist mentioning a few more. In another dream I am in a futuristic setting, with an android that has got 'out of control'. The scientist in charge, Anthony Hopkins, calms it down. The android is standing on a platform and Hopkins makes some adjustments. This is before I have watched any episodes of *West World*, of which Hopkins is a star; and when I do watch the series for the first time, I see an episode in which this happens. In a dream, Richard Tarnas, author of *The Passion of the Western Mind*, calls to tell me I am fired. In 'real life' I didn't work for him, but I did contribute to a journal, *ReVision*, that he was then editing. I did get a call from him that day, but not about that. But in the same day's post I got a rejection letter from a magazine I had sent an article to. Here is another example of symbolic distortion, and also the dreaming mind's penchant for combining different bits of 'real life': in the dream Tarnas fires me; in 'real life' I don't work for him, but I have contributed to a journal he is editing. He does call, but not about this, but I *do* receive a rejection slip from another editor, and so I *am* 'fired', or at least I am rejected. (I am happy to say that I made some changes to the article and the editor then accepted it.)

Practical value of precognitive dreams

At this point, a reader might ask, aside from the philosophical or psychological or even scientific significance of these dreams – and of course their undoubted strangeness – what practical value did they have? That is a good question. As Jung and most serious readers of dreams recognise, we dream for a reason; I have benefited in very practical ways from my precognitive dreams and from my dreams in general, although I have not always heeded them as much as I should – something sadly true of most of us. And while I have not had to avert a disaster or prevent an accident – and pray I will never have to – having dreamed some events in advance has been a benefit in more personal ways.

One such occasion once again involved my ex-wife. One afternoon, not long after moving to London, I was in the back of a bookshop and saw that F. – who had moved there as well – had walked in. We hadn't been in touch and I didn't want to see her, and so I stayed out of sight. That night I dreamed that I was walking along Parkway in Camden Town, towards Regent's Park, when I heard someone behind me call my name. It was my ex-wife. Hmm, I thought, and chalked that one up to guilt. But that evening, I *was* on Parkway, on my way to a lecture given at Kathleen Raine's Temenos Academy, which at the time was meeting in rooms on Parkway near Regent's Park. As I walked, I heard someone behind me call my name. It was my ex-wife. She was on her way to the lecture too. I wanted to avoid her but my unconscious mind knew better and had arranged a meeting, even giving me fair warning in a dream. We needed to talk and my unconscious made sure that we did.

I have avoided arguments because I recognised at the start that I had dreamed about the quarrel. This may not sound that important, but alerting us to imbalances in our attitudes is one of the values of dreams and while this may not seem as dramatic as preventing a disaster, most of the disasters in our lives come in these smaller, personal forms. Precognitive dreams have also given me signs of encouragement and the feeling that despair and depression are wasteful indulgences.

One such message came to me years ago, when I was working at a well-known metaphysical bookshop in Los Angeles. In a dream I was with a group of Tibetan monks who were gathered in a circle. They were unravelling a very large and very knotted ball of thread. I joined them and we began to dance around the circle. One of the monks handed me the ball of thread, and as we danced I began to undo it. At first it seemed complicated, but it soon became easy. And as I unravelled it, the thread widened out and the monks took hold of it. It turned into a beautiful mandala, rather as if someone had knitted it, and it expanded as we danced. When the monks saw this, they were delighted and smiled.

The next day, at the bookshop, as soon as I arrived a colleague mentioned that he had found a photograph of me in a past life. Remembering the dream, I asked, 'Was it in Tibet?' He said, 'Why, yes, in fact it was.' He then opened a copy of Alexandra David Neel's *Initiates and Initiation in Tibet*, and pointed to a photo of a round faced monk with glasses, to whom I bore a slight resemblance. When I told him about the dream my colleague was impressed and said it meant that I needed to study Tibetan Buddhism. Since then I have, but at the time I felt that the dream was a confirmation that I was on the 'right path' already. Other dreams at other times have, I believe, conveyed the same message.

Striking synchronicities

Before ending this chapter, let me risk leaving the reader even more bleary eyed by adding a few more examples. While not strictly precognitive, the synchronicities I offer here share that peculiar 'tang' that J. B. Priestley mentions. We will look at synchronicity later in the book. Right now, let us be satisfied with seeing them as 'meaningful coincidences', an experience in which something going on in one's mind seems mirrored to an extraordinary degree in something happening in the 'real world', with no obvious reason or connection to account for it, and which seems so meaningful, so *intentional*, that it leaves one wondering how on earth this could be so. This is why they are *meaningful* coincidences. It's what sets them apart from what we can call 'ordinary' coincidence, although some ordinary coincidences are so extraordinary that the distinction seems academic.

So, for example, recently I was on my way to give a talk about
Colin Wilson and his most well-known book, *The Outsider*, for the
Theosophical Society here in London. On the way, I stopped in a
local market to pick up something; I hadn't planned to, and just
thought of it on impulse. Waiting at the checkout queue, I glanced
at the magazine rack. Directly in front of me was a stack of *Vogue*,
the March 2020 issue. The magazines in front blocked everything
except the top bit of the cover. All I could see was the beginning
of the headline of a lead article. What did it say? 'The Outsider.' I
had to laugh.[16] I even took a photograph and posted it on Twitter.[17]
(It was, in fact, the kind of synchronicity that Wilson would have
enjoyed, and this one doubly so.) Oddly enough, on my way
home after the talk, I stopped in the market again; I had forgotten
something earlier. Back in the same queue, I looked for the copy
of *Vogue*, but it was gone; the issues had been changed, and the
new issue was in place of the one I had seen earlier. If I hadn't gone
into the market just then, on my way to talk about *The Outsider*, I
would not have seen the title of the article. It seemed to be waiting
for me.

Again, while going over my dream journals for this book, I
came across a dream from 1998 in which someone tells me to 'Just
stay home. There's no reason to go out. Just stay home, where it
is safe.' Not particularly interesting in itself, but this was precisely
the advice the government was giving at the start of the corona
virus pandemic and which led to the nationwide lockdown that is
still in place as I write, and which started when I came across this
dream. I tweeted about this too.[18] I would not consider this dream
precognitive, but I do think it is quite a coincidence that I came
across it while researching a book about precognitive dreams, and
that the advice I am given in it was being broadcast around the
world at the same time that I came across it.

And lastly, two more synchronicities. I have mentioned these
before but they deserve to be repeated.[19] Aptly enough, they both
involve Jung. In the first, I was with my sons and their mother in
Munich, staying at a friend's over Halloween. We had a pumpkin
to carve into a jack o' lantern. I was writing my book on Jung at
the time and earlier in the day I had watched *C. G. Jung at the
Bollingen Tower Retreat*, an excellent documentary about Jung's

stone carvings. In it, Jung tells the story of receiving a huge stone block by mistake, and explains that he kept it because it seemed a 'perfect cube.' That evening, while carving the pumpkin, my older son, who was ten at the time, pulled out a piece he had just carved. He held it up to me and said 'Look at this! A *perfect* cube!' He had never used those words before and I hadn't heard the story of Jung's cube until that day.

And finally, one last appearance by my ex-wife, who, if she reads this book, may feel that I have paid more attention to her here than I ever did during our marriage. Appropriately enough, the synchronicity in question is about precisely that. While waiting for a bus in Camden Town in London, I was reading Jung's account of a patient who had come to him for help regarding a woman he wanted to marry. He had a strong mother complex and asked Jung for advice. Jung consulted the *I Ching*, the ancient Chinese oracle, and received hexagram Kou, 'Coming to Meet'. The oracle was direct: 'One should not marry such a woman.' As I read this I regretted that I didn't have Jung and the *I Ching* around to consult about my own marriage, which ended in failure. Just as I thought this, the bus arrived and off it stepped my ex-wife. She didn't live in that neighbourhood, did not visit it often, and, as mentioned, we were not in the habit of seeing each other.

Chapter Two
The Nightly Sea Journey

Dreams and literacy. Narrative character of dreams. Gilgamesh.
Arthur Koestler's 'reculer pour mieux sauter'. *A pictorial, pre-verbal consciousness. Creativity and dreams. No one theory covers all dreams. They change as we attend to them. Are my dreams all mine? Games of the subconscious. Dreams in the ancient world. Jokes, puns, and plays on words. Freud and the Egyptians. The Greeks and dreams. Prodromal dreams. Animals and sleep paralysis. Evolutionary archetypes. Artemidorus.* Träume sind schäume. *REM. Swedenborg. Romantics and the dark side of the mind.*

Human beings have been interested in dreams for as far back as records go. As the earliest written accounts of human life show, we have wanted to know what dreams are, what they mean, where they come from and what we are to make of them, ever since our first attempts at fixing our thoughts in some external form. 'No sooner had people discovered the art of literacy,' the Jungian psychologist Anthony Stevens writes, 'than they began to record dreams.'[1]

Some of the earliest accounts of dreams date from before 3000 BC and were found in the library of the Babylonian king Ashurbanipal, who reigned from 669 to 626 BC. That a king believed that records of dreams going back some 2500 years before his time were worth preserving suggests how important dreams were for the ancients. Among the other archaeological treasures discovered when Ashurbanipal's palace was excavated in the mid-nineteenth century, were the tablets recounting *The Epic of Gilgamesh*, generally regarded as the earliest surviving work of great literature. That dreams and poetry were discovered in the same archaeological dig itself seems an example of the strange correspondences that are so much a part of dreams. Not only does *Gilgamesh* contain the first record of a dream interpretation – when

Ninsun, Gilgamesh's mother, tells him that his bad dreams herald the arrival of Enkidu – more than one dream explorer has noted the literary character of dreams. Dreams tell a story, they introduce dramatis personae, and use metaphor and analogy. More than one literary genius has been fascinated with and given us fascinating accounts of their own dreams, from Thomas De Quincey to the horror writer H. P. Lovecraft, while more than one writer or poet has revealed how dreams have inspired their work. Robert Louis Stevenson's account of how the inspiration for *The Strange Case of Dr Jekyll and Mr. Hyde* came to him in a dream is a classic example of this, as is Coleridge's account of writing 'Kubla Khan'.[2]

An archaeological excavation, a descent into our past, seems an apt start to the history of dreams, as in dreams themselves we seem to return to an earlier state of consciousness, one in which metaphor, symbol, and image are the dominant means of communication, and not the linear, sequential mode of rational, logical thought, which, in evolutionary terms, is a very recent development. As Arthur Koestler, no stranger to coincidence and synchronicities, points out, and as other dream explorers have echoed, there is a distinct similarity between creative states of mind and those that we associate with dreaming. Each night when we undergo what Jung called 'the night sea journey', we enter our creative depths.

Reculer pour mieux sauter

'In the decisive phase of the creative process,' Koestler writes, 'the rational controls are relaxed and the creative person's mind seems to regress from disciplined thinking to less specialised, more fluid ways of mentation.' Koestler continues, 'A frequent form of this is the retreat from articulate verbal thinking to vague, visual imagery.'[3] Even as rational a dream investigator as the preeminent pre-Freudian sexologist Havelock Ellis agreed. Although the penchant of dreams to 'discover analogies' is 'doubtless a tendency of primitive thought,' Ellis wrote, it is also a 'progressive tendency.'[4] Koestler characterised this regression in order to progress by the

French phrase *reculer pour mieux sauter*, to 'draw back in order to make a leap'. The leap that the mind thus regressed takes is often into some rather unusual places.

No one theory can account for the entire phenomenology of dreams. Anyone who takes the trouble to look into the history of our attempts to understand what dreams and dreaming are, and is at all honest, must reach this conclusion. And an attentive observer of dreams must as well. I know from experience. Unlike other phenomena, dreams do not always present the same face. They vary. And as Ouspensky noted, the attention we pay them alters them – or perhaps more accurately, they alter when we pay them attention, a subtle but important distinction.[5]

Often it seems that the agency responsible for our dreams is more of a 'who' than a 'what'. At the least it is the case that dreams have the same tendency to respond to observation that Werner Heisenberg discovered in the electrons whose position he could determine but whose speed he could not, at least not at the same time. The uncertainty this raised was jokingly accounted for by some physicists who remarked that the electrons knew they were being observed, and decided to play hide and seek. The same could be said for dreams, and it was, in fact, by Havelock Ellis. 'One might almost say that in dreams subconscious intelligence is playing a game with conscious intelligence,' Ellis remarked.[6] Exactly who this subconscious intelligence is, Ellis doesn't say. But it seems to enjoy a very large playing field. 'It may well be,' Ellis reflects, 'that the dream-process furnishes the key to the metaphysical and even … physical problems of our waking thoughts, and that the puzzles of the universe are questions that we ourselves unconsciously invent for ourselves to solve.'[7]

I doubt if Ellis meant his remark to be taken in this way, but this does seem to suggest that 'life is but a dream', at least in the sense that the mysteries that dreams present to us may be the same mysteries we are confronted with in our waking world.

Multiple character of dreams

Some of my dreams are Jungian, some Freudian, and some seem not mine at all, a suspicion that J. B. Priestley and the philosopher William James held towards some of their dreams – that is, that their source was in another mind.[8] This was an idea they shared with the pre-Socratic philosopher Democritus, who believed that dreams were caused by a kind of telepathy.[9] Some dreams are precognitive, some about the past (often with a detail and vividness that is astonishing), and some seem to take place in a 'timeless' zone of some kind, an insight T. C. Lethbridge explored. Some dreams are breathtakingly creative, some dull, and some are simply impossible to describe, what the philosopher of consciousness Jean Gebser called a 'kernel dream', one that is indubitably felt but impervious to explicit articulation.

Some dreams take place in seconds, some seemingly last all night, some fade away upon awakening, subject to what Koestler called *oneirolysis* (from the Greek *oneiros*, dream, and *lysis*, dissolution).[10] Some make their presence felt throughout the day, and some make such an impression on us that we remember them throughout our lives. J. B. Priestley remarked that the 'most meaningful and the most ecstatic moment' he had ever known occurred in a dream, and yet 'I have for it less visible evidence than I have for a slight cold in the head or a broken fingernail.'[11] This is a rather impressive assessment of an experience that, for many people, including many scientists, is just something going on 'in your head'. Far from being airy nothings, insubstantial vagaries of the befuddled sleeping mind, dreams have had a powerful influence on human life and destiny.

Dreams and the ancients

The Egyptians, Assyrians, Hebrews, Greeks, Chinese, Hindus, and other ancient peoples all took dreams very seriously and had a variety of ideas about what they might mean and what their source was. The Egyptians regarded dreams as warnings from the gods –

the Babylonians thought they came from spirits – and practised a strange kind of 'reverse psychology' in interpreting them. Although they generally regarded dreams as benevolent, for the Egyptians 'good' dreams tended to presage bad things and 'bad' dreams the opposite.[12] So in today's terms, if you dreamed that you won the lottery, it might mean that you would lose money in the near future.

The Egyptians seem to be the first people to practise dream incubation, at least according to the historical record; one can't help but suspect that people practised some form of this before records began.[13] They used herbs, drugs, and spells to attract beneficial dreams and to ward off evil ones, a practice still popular today with contemporary dreamers, Egyptian or not.[14] Temples to Serapis, the god of dreams, were dotted throughout Egypt; the one at Memphis, the most famous, dated to around 3000 BC. The Egyptians put questions to the gods and sought their answers to them in their dreams, and their methods of interpretation, which can be found in what is known as the Chester Beatty papyrus, dating from around 1350 BC, are strikingly similar to our own modern understanding of them.

Freud and the Egyptians

As Anthony Stevens points out, much of how the Egyptians saw dreams is similar to how Freud saw them. As did Freud, the Egyptians looked for hidden associations and correspondences in dreams, sought out contraries in understanding them – their 'reverse psychology' approach – and recognised that one of the habits of dreams was to make puns and tell jokes.[15] Although dreaming is a serious business, the Egyptians, as well as others who studied their dreams, saw that whatever agency is responsible for them, it has a sense of humour. The dream artist who can recreate reality better than any effort of photorealism and who can parade before us stunning scenery and involve us in incredible adventures, even mystical experiences, has, it seems, a weakness for witticisms. Havelock Ellis recognised this when he wrote that 'The conditions of psychic activity during sleep seem to be specially favourable to the productions of puns and allied forms of witticisms.'[16]

But, as Stevens points out, what might have been funny to an ancient Egyptian most likely won't raise a laugh today. Puns, plays on words, and jokes are possible within a cultural and linguistic context that is peculiar to the dreamer and his contemporaries. This emphasises the fact that while dreams speak in a pictorial language made of symbols and metaphors, these are most often immediately related to the dreamer's life. The dream is a story about you – or at least it can be, most of the time.

One consequence of this is that most 'dream interpretation' books, telling you what it means if you dream about a cat or of your teeth falling out, are worthless, as the important question is what a cat or losing teeth mean in *your* life, not in the abstract, or to an ancient Egyptian, for that matter. We can think of dreams as a game of charades in which it is up to you to guess what the dream is aiming at. 'Big', archetypal dreams, in which the symbols are universal, welling up, as Jung says, from the 'collective unconscious', are, of course, part of our dream life. But these are few and far between, at least for most of us, and we know we have had them by their numinous character. These tend to be the unforgettable dreams that stay with us throughout our life.

But most of our dreams relate to our everyday lives, our problems and challenges, and a joke a dream may make is a way of communicating some knowledge helpful in this. Or, as is most often the case, a pun in a dream is a way of sugar coating some pill of self-knowledge that would otherwise be difficult to swallow.

Dreams and the Bible

Mention of Freud reminds us that the Jews were considered the dream experts of the ancient world. Probably the two most well-known examples of dream interpretation can be found in the Old Testament, Joseph's interpretation of Pharaoh's dream of seven fat and seven lean cows, and Daniel's reading of Nebuchadnezzar's dream, which precipitated a breakdown.[17] Whole chapters of the Talmud are devoted to dreams, and prophetic dreams are a common theme throughout the Bible. One of the most quoted remarks about dreams is credited to Rabbi Hisda, a fourth century Talmud scholar. 'A dream uninterpreted,' Rabbi Hisda said, 'is like a letter unread.' His contemporary, Rabbi Bizna, argued that a single image or symbol in a dream can have multiple meanings, a semantic condensation that has kept many a dream reader well employed. Other people of the book have been just as interested in dreams. Joseph is told of Mary's miraculous pregnancy in a dream, and he is warned to flee Herod's massacre in the same way. And Islam is founded on a dream, Muhammad's dream 'night journey', the *Lailat al-Miraj*, in which he was accompanied by the angel Gabriel on the half-human horse Elboraq on a miraculous flight and 'initiation into the mysteries of the cosmos'.[18]

A butterfly's dream?

Maya, the mother of Gautama Buddha, was told of her pregnancy in a dream, as Joseph was told of Mary's. And one of the most well-known dreams is that of the Taoist philosopher Chuang Tzu, who lived in the fourth century BC. In the *Zhuangzi*, Chuang Tzu wrote that he once woke from a dream in which he was a butterfly, and knew nothing of his life as Chuang Tzu. Now, wide awake, he wonders if what he is experiencing is but a dream of the butterfly, thinking it is Chuang Tzu. The strange metaphysical character of some dreams and their potential to undermine our sense of identity – a concern made much use of by paranoid science-fiction writers such as Philip K. Dick – has troubled serious dreamers ever since. William James was so troubled by a dream that he woke from it unsure of his identity and wondering if he was 'getting into other people's dreams.' It was, he said, 'the most intensely peculiar experience of my whole life.'[19]

René Descartes, one of the architects of the modern rational world, was troubled by the seeming reality of dreams and was led to his famous formula, *cogito ergo sum* – he thought, therefore he was – by way of reflecting on how strangely our dreams seem to duplicate our waking world, and how difficult it was to tell the two apart. That the ancient Hindus were well aware of this difficulty can be seen in the taxonomy of dreams devised by the authors of the *Vedas*, who tended to see dreams as lucky or unlucky, rather than good or evil. In the *Artharva Veda*, compiled circa 1200 BC, we are told that dreams take place in an intermediary realm, situated between the two worlds, that of the physical plane and that of the higher, spiritual ones.[20] In Tibetan Buddhism, this distinction between 'real' world and 'dream' world is dissolved in the recognition that both are illusions, a conclusion that the practices of Tibetan 'dream yoga' aim to instil in their devotees. Through developing the power of retaining wakefulness in sleep – what we today call 'lucid dreaming' – the practitioner comes to see the illusory character of all existence, and enters what is known as *turiya*, a state of deep, dreamless sleep – or wakefulness.

The Greeks and dreams

The beginning of what we can call a rational, scientific approach to dreams seems to have appeared with the Greeks. To be sure, they too believed in dreams as visitations or messages from the gods. The temples to Asclepius, the god of medicine, scattered over ancient Greece, wherein visitors hoped to receive a redeeming dream, tell us that the Greeks followed on from the Egyptians in the practice of incubating dreams.

The ancient dream healers at Epidarus and other sites sacred to Asclepius emphasised the importance of 'set and setting' – as Timothy Leary would also, millennia later. To reach the temple itself, which was in a place of natural beauty, was something of an ordeal, and before the patient could avail himself of what he hoped would be a revelatory and restoring sleep, he had to undergo a ritual purification. His clothing was taken; he bathed, fasted, and was given a white robe and led to his place within the sacred space, where he intoned his prayers to the god. Then he was most likely given a libation to help his sleep, probably a draught containing opium, long known as a bringer of dreams. Sometimes the dream would communicate some message, some information that would help relieve the patient's ailment. Sometime the dream alone, within which the god Asclepius himself would often appear, was enough to effect a cure.

Today it is still not uncommon for a powerful dream to have a rejuvenating or curative effect, one from which the dreamer awakes, knowing that something deeply significant in themselves has changed. Such a dream comes to the reprobate Dmitri Karamazov in Dostoyevsky's *The Brothers Karamazov*. When he wakes from it, he is willing to accept the punishment for his father's murder, even though he has not committed it, because he knows the suffering it will entail will *change* him. 'I have had a good dream, gentlemen,' he tells his accusers, 'with a new light, as of joy, in his face.'[21]

Not all dreams were the same for the Greeks, who turned a critical eye on many things their neighbours took for granted. In the *Odyssey*, Homer has Penelope dream of an eagle swooping down and killing her pet geese, which is meant to symbolise Odysseus' return and routing of her suitors. As she tells this to Odysseus – who is in disguise – she remarks that while she knows what the dream means, she is concerned that it may be a false dream. She tells Odysseus that the dreams that come through the 'gate of horn' are true, but those that arrive through the 'gate of ivory' are false, the distinction between the two turning on a play on words in ancient Greek that cannot be captured in English.[22] How to determine a true dream from a false one, a prophetic dream from one born of wish-fulfilment, is a conundrum contemporary dreamers still face. As it was in ancient Greece, it still is today: there is no hard and fast rule for this, and the dreamer must learn through the most demanding school of all, experience.

The sleep of Logos

One of the first Greek sages to remark on dreams was Heraclitus, the Ionian pre-Socratic philosopher who lived in the sixth century BC. For him, dreams take place in an entirely subjective world. When we dream, Heraclitus said, we are cut off from the sunlit, objective world of reality, and are submerged in a purely private experience. The gods have nothing to do with it; what's worse, in dreams we are cut off from the logos or 'reason' that runs through the cosmos. Heraclitus seems not to have experienced shared or mutual dreams, but that aside, we know what he means. Who hasn't surfaced from a nightmare utterly relieved to find that it was 'only a dream' and thankful for reality? Heraclitus' belief that dreams are less real than our daylight reality and are thus of much less importance and should be ignored, is more or less how many, if not most people in the modern world view dreams. In fact, Francis Crick, co-discoverer of the DNA molecule, has gone on record advising that dreamers should not try to remember their dreams. For Crick, dreams are psychic waste matter, and so trying to recall them is, in a way, rather like being psychically anal retentive.[23] (It is somewhat encouraging to know that I have contravened Crick's advice for the past forty years with, so far, no ill effects.)

Yet Heraclitus' remark can also be taken to mean that most of us continue to live in our private, subjective worlds, even when awake, an insight he shared with the esoteric teacher Gurdjieff. The task of the philosopher is to awaken, first himself, then others, a challenge Plato continued to address some centuries after Heraclitus, when he compared human existence to life in a cave in which all we see are shadows on a wall. Plato himself believed that in dreams we act out desires and frustrations that in conscious life we cannot, an anticipation of Freud's idea of dreams as fulfilling repressed wishes, a few millennia ahead of psychoanalysis. I should point out that Socrates, Plato's teacher, and the founder of the Socratic method of critical inquiry, had a repeating dream in which he was told to 'make music'. According to Jung, this was an example of the unconscious gently suggesting to Socrates that he had to compensate for his ardently rational vision with a bit of poetry.[24]

Prodromal dreams

Hippocrates, the 'father of medicine' – doctors today still take the Hippocratic oath – lived a few centuries after Heraclitus.

He believed that in his patients' dreams he could often detect the early signs of an approaching illness; such dreams have come to be called 'prodromal', which means the period from the first indications of a disease to the outbreak of its full symptoms. The great Galen, the second century Roman physician, wrote of such dreams in his *Prophecy in Dreams*. They are still being used today: in *Awakenings* Oliver Sacks told of how people suffering from encephalitic lethargica ('sleeping sickness') dreamed of their illness just at its outset.

Though not strictly precognitive, these dreams are signs that the sleeping mind is alert to changes in the body that have not yet risen to consciousness, and can anticipate the approaching illness in the form of images or symbols that the physician learns to interpret. This is not altogether different from how a psychoanalyst operates; lying on the coach, the patient tells the psychoanalyst their dreams, which the analyst scrutinises for signs of emotional or psychic disturbance. As many psychotherapists know, a dream can announce a psychotic episode in advance of its arrival.

Aristotelian dreams

The shift to what we would recognise as a 'scientific' approach to dreams seems to have begun with Aristotle, which should not surprise us, as Aristotle is considered the first 'research scientist'. Unlike his teacher Plato, who was not particularly interested in the natural world, Aristotle stuck to observation and collected evidence before advancing theories. For Aristotle, who wrote extensively on dreams, dreams are a purely natural phenomena, produced by fluctuations in our bodily states. In *On Dreams* he even suggests indigestion as the cause of some dreams, a classical source for a common dream explanation.[25]

This was an assessment Aristotle shared with Havelock Ellis and, oddly enough, Ouspensky, although Ouspensky did recognise the reality of 'big dreams', ones that originate 'in the innermost recesses of life and rise above the common level of our understanding and perceptions of things' and 'can disclose a great deal that is unknown to us', such as the future.[26] Aristotle himself thought precognitive dreams could be explained by coincidence. We dream of so many things, he thought, that it stands to reason we would every now and then dream of an event before it happens; again, this is the usual explanation for future dreams. And Aristotle also recognised a fact that has informed much of modern scientific investigation of dreams: that animals dream too.

Anyone who watches their dog growling and scraping its paws as if in pursuit while it is curled up by the fire can observe what Aristotle noticed millennia ago. Animals seem to 'act out' their dreams, and would truly go through the same motions as their dream selves perform, were it not for a knack they picked up since dreaming first entered the evolutionary story more than 130 million years ago.[27] When we dream, our brain shuts down the centres controlling our movements; this was an evolutionary trick our animal ancestors learned so that they would stay still and quiet during sleep and not attract predators. This suggests that dreaming was something important enough to preserve; if not, it could have been jettisoned altogether along with other characteristics that did not make the evolutionary grade. This is an important point we should not lose sight of: dreaming is so much a part of our life that evolution itself took steps to protect it.

Sleep paralysis

When we dream, our muscles become torpid, that is, slack and unresponsive, and this is the physiological explanation for an experience that I have had on many occasions, what is known as 'sleep paralysis'. Sleep paralysis happens when for some reason you – that is your conscious self – 'wakes up' but your body is still asleep. It can be very disturbing and over the centuries sleep paralysis has been associated with visitations by incubi and succubi, 'out-of-body' experiences, and alien abductions, among other things.

That mammals dream – reptiles and amphibians seem not to – and that dreaming itself arose during the Cretaceous period, seems to undermine some of Freud's ideas about dreams, unless we want to assume that our animal predecessors suffered from Oedipus complexes well before Sophocles thought of writing *Oedipus Rex*. And oddly enough, the fact that animals dream seems to support some of Jung's ideas about 'archetypes' and the 'collective unconscious', at least from the perspective of evolutionary biology. This is a neat twist, as Jung has often been criticised by Freudians for being 'unscientific', while today Freud's ideas are considered by many to be a form of 'pseudo-science'.[28]

Evolutionary archetypes

Put briefly, Jung's archetypes, the psychic 'blueprints' for experience that we inherit from the collective unconscious, have acquired a new lease on life in the form of 'innate release mechanisms', programs in ourselves and other animals that activate instinctual behaviour when faced with the appropriate demand. Yet what has been observed is that these 'releasers' are ready and waiting in the psyche of an animal (and ourselves), even if the external factors generally regarded as responsible for them coming into existence, are never experienced by the animal.

If you know your neo-Darwinism, this should be impossible. It tells us that the behaviour that animals – ourselves included – exhibit is learned; it is a response to the impact of the environment. This is an expression of the general *tabula rasa* ('blank slate') picture of the mind, human or animal, maintained by orthodox science. This agrees with John Locke's axiom, announced in his *An Essay Concerning Human* Understanding, which first appeared in 1689, that 'there is nothing in the mind that did not get there by way of the senses.'[29]

Yet what seems to be the case is that some behaviour comes *before* the animal has had any experience of the external factor that triggers it, rather like the software a computer comes with, which is ready to use whether you ever do use it or not. Chicks respond to a cardboard image of a hawk swooping over them, long before they have had enough time and experience to learn to avoid being eaten or to ever have seen a 'real' hawk. In fact, they would have to, as those without these inherited programs would not survive. And what seems even more remarkable, is that some animals respond to releasers that are *not* found in nature, but to exaggerated versions of the 'real thing'. As Colin Wilson remarks, talking about the research into 'super-normal releasers' by the psychologist Stan Gooch, 'The ringed plover responds more strongly to white eggs with black spots than to its normal light brown eggs with darker brown spots.'[30] The same is true of other animals.

Gooch, in his important book *Personality and Evolution*, suggests that something along these lines may be at work in 'sexy' men's magazines, with photographs of women with exaggerated breasts and bottoms that trigger their readers' release mechanisms more quickly and powerfully than anything they might encounter 'naturally'.[31] This gives new meaning to the idea of the 'dream woman' many men seek throughout their life (what Jung called the 'anima') and, more often than not, do not find. What it means in terms of Jungian psychology is that if archetypes and innate release mechanisms are related, then archetypes may have evolved through natural selection.[32]

Artemidorus

The Greek all dreamers and those interested in dreams owe a debt of thanks to is Artemidorus, a physician of the second century AD who hailed from Ephesus in Asia Minor, an important city known for its Temple to Artemis, one of the 'seven wonders of the ancient world'. Artemidorus travelled around the then known world – which meant the Roman Empire – collecting accounts of dreams, speaking with dream interpreters, soothsayers, and physicians, and scouring libraries. After years of diligently gathering evidence, he wrote his *Oneirocritica*, a five-volume work on the interpretation of dreams. Freud thought so highly of his ancient predecessor that he borrowed the title of Artemidorus' work for that of his own. *Oneirocritica* means 'the interpretation of dreams'. Freud's own *The Interpretation of Dreams* was published in 1899, and throughout it he carries on a conversation with the individual who is credited with writing the first 'dream book'. As you might suspect, they often arrive at different conclusions.

Yet unlike most dream books to come, Artemidorus rejected the kind of 'one size fits all' approach that many popular dream interpretations take. The meaning of symbols can change over time, even with location, Artemidorus recognised. And, as mentioned earlier, what was of most importance was what part the symbol played in the dreamer's own life. (And we should remember that, in a dream, *anything* can be a symbol.)

The dreamer's personal circumstances were crucial and, taking a cue from the Egyptians, Artemidorus cautioned that one should always be on the lookout for puns and word play. Artemidorus was said to have studied more than 3000 dreams so we might be inclined to listen to what he had to say.[33] As Anthony Stevens points out, Artemidorus took an empirical approach, studying each dream as a new, fresh phenomenon, and like Freud and Jung he was particularly keen on the associations the dreamer had with the images of his dream, and those that he, the dream reader, had too. Like Jung – but unlike Freud and Aristotle – he did believe in prophetic dreams, and he made a distinction between 'little' dreams, which he called *insomnium*, which dealt with everyday life, and 'big' dreams, which he called *somnium*, or 'true' dreams, which concerned larger, deeper issues. Here we can see an early distinction between the dreams of the personal unconscious – to which Freud limited practically all dreams – and those emerging from what Jung called the collective unconscious.[34]

Before leaving our brief look at these ancient oneironauts we should at least mention one of the early church fathers, Synesius of Cyrene, who lived a century after Artemidorus. Like Artemidorus, he believed that in dreams we can catch a glimpse of what lay ahead. The faculty involved in this was what he called 'fantasy' and which we would understand as 'imagination'. Through the imagination, Synesius believed that potentials and possibilities dormant in the soul can be brought to awareness. After that it was the task of the conscious mind to make them realities. This is an idea we shall return to.

Science and dreams

Perhaps at this point it might be good to take a look at how mainstream science today understands dreams. This should not take long, as the general assessment from this quarter is along the lines of Francis Crick's above; that is, that dreams are meaningless. That many scientists have acknowledged the importance of dreams in their own life, and have credited them with many inventions and discoveries, seems not to have made much difference to this view. As Brian Inglis remarks, 'The influence of dreams on science has been airbrushed out of most biographies of scientists.'[35] The list of 'dream discoveries', though, is impressive.

Probably the most famous dream discovery is that of the nineteenth century German organic chemist August Kekulé, whose dream, or hypnagogic vision, of atoms forming chains that became snakes swallowing their own tails – the ouroboros, an alchemical symbol and also that of the 'eternal recurrence' – led to his establishing the ring structure of the benzene molecule. The mathematician Henri Poincaré was led to the discovery of a class of Fuchsian functions through a similar hypnagogic experience.[36] But, as Inglis points out, Kekulé and Poincaré were not alone.

In 1869, the Russian scientist Dmitri Mendeleev dreamed the periodic table of the elements. The physicist Niels Bohr's discovery of the structure of the atom, with the electrons whirling around the nucleus like the planets round the sun, for which he won the Nobel Prize in 1922, came to him in a dream. Elias Howe gave credit for his invention of the sewing machine to a dream. Insights into the world of insects came to the great French naturalist Henri Fabre in a dream. Thomas Edison, to whom we owe the lightbulb, among other things, made use of the hypnagogic state to inspire new ideas. Edison used to nap while holding ball bearings in his hand; as he drifted off, the bearings would drop into a steel pan and he would be jolted awake, usually with a new idea or a new approach to a problem that had him stumped.[37] I've mentioned Descartes already and the difficulty of differentiating between dream and reality. But a dream Descartes had in 1619 when he 'was filled with enthusiasm' about his discovery of 'the foundations of the marvellous science' – that of applying mathematics to philosophical problems, which would radically transform human life – was for him the most important event of his life.[38]

In a famous address during which he told the story of the origin of the benzene ring, Kekulé urged his audience to follow his example. 'Let us learn to dream, gentlemen,' he told his learned colleagues. Many had already done so, and others would too. One scientist who thought very highly of dreams was the physicist and Nobel Prize winner Wolfgang Pauli, discoverer of the neutrino and the Pauli Exclusion Principle, who brought many of his dreams to Jung while in analysis with him.[39] Jung later used Pauli's dreams as the basis for his long essay 'Individual Dream Symbolism in Relation to Alchemy', which makes up part of his late work *Psychology and Alchemy*.[40] For Jung, Pauli's dreams reveal the process of 'individuation', how 'one becomes who one is'. This meant becoming psychically whole, a goal, Jung believed, that the alchemists pursued in their search for the philosopher's stone. Most physicists consider alchemy as absurd and as meaningless as dreams; not Pauli. Pauli and Jung later collaborated on a joint work about synchronicity, to which Pauli himself was peculiarly subject.

Träume sind Schäume

Here I've given only a few examples of how dreams have informed not only poetry and myth, which we might expect, but also technology and science. I could give many more – Brian Inglis' *The Power of Dreams* is full of them – but the reader, I think, gets the idea. Yet with all this, the standard appreciation of dreams from the orthodox scientific view – although it is difficult to get more orthodox than Nobel Prize winners – is that, in the words of a German proverb, *Träume sind Schäume*, 'dreams are foam', a kind of psychic froth churned up by the brain at night. The German physician Adolf Strumpell, an early reader of Freud, said that dreams were like 'fingers wandering over piano keys', a kind of mental noodling. We've already seen what Francis Crick thought of them. His assessment is on a par with that of J. Allan Hobson, one of the most prestigious of dream scientists. For

Hobson, author of *The Dreaming Brain*, dreams are 'senseless, random accompaniments' of the electrical activity in the sleeping mind, unchained, as it is, from constraints of the sensory world that otherwise gives it direction. In *Landscapes of the Night: How and Why We Dream*, the computer scientist Christopher Evans not surprisingly adopted a computer model to explain dreams. He saw them as a product of the brain's being 'off-line', a kind of random, meaningless activity going on while it processes the day's 'input' and 'updates' its apps for the next day.

Not all scientists interested in dreams see them in such reductive fashion. Unfortunately, those who do not are often accused of doing bad science by those who do. Some scientific views of dreams ignore what they may mean and focus on what they *do*. They are not messages from the gods or from our unconscious, but dry runs for life. From this perspective dreams are a way for us to rehearse actions that we will have to perform in life; in this sense they relate to the idea of archetypes as innate releasing mechanisms, with the archetypes preparing us for the world we will have to face. This is a practical, utilitarian approach to dreams that seems to make much sense – until we reflect that some dreams are so bizarre and complex that it is difficult to see exactly what events and situations in life they may be rehearsing us for.[41] Or why, say, having had a sufficient number of dreams to prepare us for some life challenge, we continue to have these dreams, even after we can be expected to have 'got it'.

REM

Yet all dreamers owe a debt of thanks to Eugene Aserinsky and Nathaniel Kleitman of the University of Chicago who in 1953 established that REM sleep is associated with dreaming.[42] REM stands for 'rapid eye movement', the apparent movement of the eyes under the eyelids while we are asleep and dreaming. One of the lyrics to my song 'Presence, Dear' is 'I stay awake at night and count your REMs' and as Anthony Stevens points out, our eyes move while asleep, as if we were watching something 'for real', is a fact that was noticed by the ancients.[43] Aserinsky and Kleitman discovered that, if woken up during a phase of 'rapid eye movements', a sleeper invariably will report that they had been dreaming, even if under normal conditions they might say that they never dream.

One of the odd things about REM sleep is that it is quite close to our waking state; this is why it is often called 'paradoxical sleep'. In many ways, for all intents and purposes, our brain is just as awake in REM sleep as it is during the day, which again suggests that if dreaming is like waking, then our waking life is like a dream (or, as Gurdjieff said, we are all asleep).[44] This is one reason why it is more difficult to wake someone from REM sleep than it is from a sleep when they are not dreaming, a finding that confirms the observation of the nineteenth century French dream investigator the Marquis d'Hervey de Saint-Denys, an early explorer of lucid dreams. In his classic work *Dreams and How to Guide Them*, Saint-Denys wrote that 'I have invariably observed that the more vivid my dreams, the greater my difficulty in waking.'[45]

Along with these rapid eye movements are other characteristics associated with this state, which is 'quite different from any other psychophysical state'.[46] Breathing becomes quick and irregular, and the brain, as mentioned, is as active as it is when awake, emitting low voltage patterns of electrical activity. It is in REM sleep that our muscles go slack and torpid, creating the 'sleep paralysis' that prevents us from acting out our dreams. One part of our anatomy, however, is not paralysed. In REM sleep, men experience erections and women increased vaginal secretions. This nod to Freud's belief that sex and dreams are inextricably mixed makes it possible for us to act out at least some behaviour in dreams: in 'wet dreams' men and women can experience orgasms that are more powerful than those achieved normally. I should point out that with all due respect to Freud, these physiological changes associated with sex happen in REM sleep whatever we dream about.[47] But then Freud would say '*Naturlich*', given, from his perspective, that whatever we dream about, it's about sex.

Ups and downs

Our descent into sleep and our entry into the land of dreams follows a particular route. When we lie down at night, we first drift into the hypnagogic state. Then, as we 'fall' deeper into sleep we enter a stage of non-REM (NREM) sleep associated with slow brain waves that lasts for about ninety minutes.[48] We then 'surface' a bit and enter REM sleep; our brain waves become quicker, our heartbeat and breathing speed up and become irregular, and we start to dream. This first phase of 'paradoxical sleep' lasts from ten to fifteen minutes. Then we sink back down to NREM sleep. There are intermediary stages between NREM and REM sleep, characterised by brain wave patterns known as 'sleep spindles' and 'K-complexes', but the basic 'movement' is 'down' into NREM and then 'up' into REM sleep.[49] This happens several times throughout the night, with the time spent in NREM sleep shortening and that in REM sleep getting longer, with shorter intervals between, and the last stage lasting from thirty to forty minutes. Then, passing through the 'hypnopompic state' – the hypnagogic state, but in reverse – we wake up.[50]

One thing that became clear through studying human sleep patterns is something that most people attentive to their dreams already knew: that there are deeper and more shallow dreams. Although the dreams we usually remember are those that appear in REM sleep, it was found that in NREM sleep the brain is still active, but what it is engaged in is a kind of thinking, not 'true' dreaming.[51] What goes on in NREM sleep is a kind of half-conscious mulling over of the day's affairs or problems, something that T. C. Lethbridge recognised in his distinction between 'true dreams' and 'drowsy thinking', a kind of 'slipshod form of re-thinking recent events and problems'. Compared to this, the 'true dream' is 'like a sudden beam of bright light'.[52]

We all know the difference between the two. We all have mornings when we wake up feeling exhausted, even after the recommended eight hours, and others when, with only a few hours' sleep, we feel refreshed. The difference may depend on how much 'true dreaming' we enjoyed in the night.

Knowledge and experience

This brief look at what takes place in the brain and body as we head to dreamland gives us an idea of the physical processes involved in sleep. But what does it tell us about our actual dreams themselves? According to some dreamers, not much. 'Apart from defining the onset and termination of the dream,' write David Coxhead and Susan Hiller, 'the biology of dreaming does not help us understand the experience.' Something that could be said about the clinical investigation of paranormal phenomena as well.[53] This may sound too extreme, but is it? Knowing the electrical processes involved in putting a picture on my television screen is very important knowledge. But it doesn't tell me much about the programmes I watch, just as understanding what the brain is doing as I write these words cannot tell me much about what they mean or what I am trying to say. We can say that there is the biology of dreaming, and there is the dream. Both are important and should be studied. But the one is not the other. Knowing what happens when we dream is important. But so is knowing the dream. Yet the distinction between the two has not always been maintained. And the value of knowing what happens when we dream has often obscured that of knowing what the dream itself is about.

As I point out in *Lost Knowledge of the Imagination*, a radical shift in how we understand the world, how we gain knowledge of it and what that knowledge means, took place in the early seventeenth century.[54] What happened then was the beginning of what we know as science. And as the popularity of what became known as the 'scientific' way of understanding our experience spread – reaching the near total dominance it enjoys today – the prestige of dreams dwindled. Among the uneducated and superstitious, of course, old beliefs die hard. But for the rational children of the Enlightenment, dreams were fantasies, at best entertaining, at worst a nightly madness – and, for the religiously inclined, often a source of guilt and shame and the conviction of sin. Yet not all enlightened minds were eager to ignore the revels of the night, nor anxious to deny them any significance or meaning. As Henri Ellenberger shows in *The Discovery of the Unconscious*, although Freud gets the popular credit for putting dreams back on the intellectual map, he neither 'discovered' the unconscious nor was he alone in his interest in its activities.[55] There were many who paved the way before him.

Swedenborg

One bright light of the late Enlightenment, the German scientist and aphorist Georg Christoph Lichtenberg, anticipated much contemporary dream research when he wrote that 'we live and feel as much dreaming as waking and are the one as much as the other. It is one of the superiorities of man that he dreams, and knows it.' And Lichtenberg added: 'We have hardly made the right use of this.'[56] But it was Lichtenberg's older contemporary, the Swedish scientist and visionary Emanuel Swedenborg, whose writings on dreams anticipated much of what Freud and Jung would rediscover more than a century after his death. Swedenborg kept a *Journal of Dreams* and in it one can find examples of the kinds of dreams that many psychotherapists encounter everyday – along, I should add, with some prophetic and precognitive ones. As Jung would, Swedenborg recognised that dreams 'tend to some definite end'. They were not random fantasies, triggered by bodily states, but 'were directed by one of whose origin we are ignorant', an early recognition of what we today call the unconscious.[57] As J. B. Priestly recognised, Swedenborg saw that dreams are 'definite creations just as a produced play or film is a definite creation.'[58]

Swedenborg's unconscious got up to some racy business. Some of Swedenborg's dreams were considered so scandalous that much of his dream journals were censored by followers eager to maintain his good reputation.[59] The most offensive entries dealt with Swedenborg's sexual dreams, in which he encountered the *vagina dentata*, a vagina with teeth, a fairly common motif in the psychoanalytical literature. Yet most of Swedenborg's dreams dealt with his scientific pursuits and, from a Jungian perspective, attempts by his unconscious – or soul, as he would say – to wean him away from his intense rationalism, something we saw in the case of Socrates. Before becoming the religious visionary he is known as today – insofar as people know him at all – Swedenborg was one of the most prolific and brilliant scientists of his time; his last major project was discovering 'the seat of the soul', which he believed, as Descartes did, resided in the pineal gland, that still mysterious organ which occupies the space Hindu philosophy reserves for the 'third eye'. As Wilson Van Dusen, an important dream explorer, points out, Swedenborg's understanding of himself and his work 'was shifted by this inward exploration', his interest in dreams. This shift was from an intensely 'objective' – that is, scientific – sensibility, to one that was 'affectional' and 'living'.[60] His dreams, we can say, were trying to awaken his feeling side.

Romantic dreamers

Swedenborg's dream journals were not published until the mid-nineteenth century. But by this time the reaction to the excesses of Enlightenment rationalism known as Romanticism had burst upon the scene, bringing with it all the dark, obscure, and uncanny facets of the human mind that the lucid thinkers of the eighteenth century believed they had forever banished. To speak of a flood tide of the unconscious would not be an exaggeration. The blood and thunder of the Gothic marked the change – William Beckford, author of the Gothic classic *Vathek*, recorded and published his dreams – and by the early nineteenth century, writers, poets, and philosophers were once again obsessed with their visions of the night.[61]

To catalogue the contributions to 'oneirology' made by these individuals would be tedious; interested readers can find the details in Ellenberger's masterwork, and in an earlier, less exhaustive work, *The Unconscious Before Freud*, by Lancelot Law Whyte, published in 1960. Goethe, Schiller, Schelling, Novalis ('When we dream that we dream,' Novalis wrote, 'we are close to awakening,' an early recognition of lucid dreaming), Hoffmann, Blake, Wordsworth, Coleridge are only some of the most well-known names associated with this movement, the central aim of which was to regain the human mind's 'other half', what we can call 'the dark side of the mind', which had been declared off-limits by the champions of rationality.

Late Romantics like Nietzsche anticipated what, in different ways, Freud and Jung would come to see. As Nietzsche wrote in *Human, All Too Human*, 'in sleep and dreams we repeat once again the curriculum of earlier mankind.'[62] An earlier thinker who set much store by dreams was Gotthilf Heinrich von Schubert, whose *The Symbolism of Dreams*, published in 1814, was one of the most famous books of the time. It was read by Goethe and other Romantics, and later by Freud and Jung.[63] Schubert was interested in what we would call parapsychology, and many of his ideas can be found in a forgotten classic of the period, Catherine Crowe's *The Night-Side of Nature*, first published in 1848 to great success (and repeated printings) and still very readable today. According to Colin Wilson, Crowe's book was the 'first sustained attempt to treat paranormal phenomena in the scientific spirit that would later characterise the Society for Psychical Research.'[64] And among the many paranormal phenomena she investigates are, as we might suspect, precognitive dreams.

Chapter Three
It's About Time

Time is…? Inability to grab hold of time, yet it is inescapable. Time and the philosophers. Parmenides and Heraclitus. Permanence and change. St. Augustine and the unaskable question. Whitehead. Does time flow? Sundials and death. Where are the banks on the River Time? Time's direction. Einstein's space-time continuum. Leibniz. Newton. The Conquest of Time. 'Be drunk!' Sacred time. The Hopi. Ouspensky's days. An eternal present. What's your hurry? Bergson's duration. Slowing down time. Watches and clouds. Our two brains. Zeno's paradox. Right brain stasis. Aldous Huxley's dishes. An earlier form of consciousness.

When we move from dreams to time, leaving one mystery for another, we enter an area of human experience that invariably slips away from us no matter how firmly we try to grasp it. In fact, the harder we try to hold on to it, the further it recedes from us, until we are left befuddled by our attempt. Then, as is usually the case, we shrug our shoulders and forget about it. The inability to do this is the sure sign that you are a philosopher.

At least we are sure that dreams exist. Except for a positivist philosopher I read years ago – and whose name escapes me now – who argued that we have no actual evidence of dreams, only of our waking up *believing* we have had them, no one, I think, denies that we dream. Our dreams may be meaningless, mental rubbish cleared out to make space for what we will pick up the next day, and all arguments to the contrary muddled, but at least we know when we have had one.

Time is a different story. With its other fundamental given, space, time seems to be one of those elements of our experience that is so palpably 'there' that to deny it or question it in any way, strikes us as absurd. In fact, as the philosopher Immanuel Kant pointed out, without time and space we would not have any experience at all, given that they are, for Kant at least, the inescapable conditions through which we experience anything. It is easy to see what Kant means. Whether we are Kantians or not, if we try to think of some kind of existence without time or space, we draw a blank. The mind boggles at the attempt, so we give it up and get back to serious business.

So, like death and taxes, time is something we can't avoid. Yet when we try to grab hold of it, it slips through our metaphorical fingers like the sand in an hourglass. That sand may tell the time for us, but it is of no help in telling us what time is in itself, what the something *is* that it, the hourglass, is measuring. And neither are the clocks which have taken over from the hourglass.

Time, then, is fleeting in two senses: it waits for no man, and flies, whether we are having fun or not, and it seems not to comply with our attempts to understand it, dissolving, like fairy gold, when we try to lay hold of it.

Yet, when we look at the clock and see that we have spent quite a bit of time pondering on the nature of time, only to have it disappear under our mental gaze, we are once again faced with its stubborn persistence. Like a dream that dissolves upon awakening, yet which affects our moods throughout the day, when we look for it, time is nowhere to be found. Yet when we look away, it is right there in front of us.

Time and the philosophers

We may feel that, not being philosophers, the fact that we can't lay hold of time should not trouble us. Yet philosophers are just as baffled as we are. According to one authority, philosophers that have 'grappled with the problem of time' have 'ended up in perplexity'.[1] One could say that for every philosopher who denies the reality of time – and there have been many – there is one who affirms it just as insistently.[2] We can even place the start of the debate in the earliest years of philosophy, among the pre-Socratics, with Parmenides arguing that change, and hence time, is an illusion, and Heraclitus contradicting Parmenides and arguing that *panta rhei*, 'everything flows'. (Plato tried to settle the debate with a compromise: in the *Timaeus* he tells us that 'time is the moving image of eternity'.) One of the most quoted remarks about time comes from St Augustine. When asked what time is, he replied 'If I am not asked the question, I know the answer.' I might mention that St Augustine also pondered the mystery of dreams, and hoped that God would not hold him responsible for his, as they were beyond his control.[3]

St Augustine experienced what we all do when we contemplate time. Before directing our attention to it, we know full well what it is, and are as sure of it as we are of the ground under our feet. It is what we lose track of and what makes us hurry when we realise we will be late for an appointment. But when we try to *say* exactly what it is, we are stopped in our tracks, and even the sturdy bedrock on which we stand seems to give way.

Whitehead and time

One might think that in the intervening centuries between St Augustine and ourselves, we might have gained a bit of traction on the 'time question'. Yet Alfred North Whitehead, a twentieth century philosopher who thought quite a bit about time, and who wrote difficult books attempting to understand it, was as baffled by it as St Augustine was. 'It is impossible to meditate on time and the creative passage of nature,' Whitehead wrote, 'without an overwhelming emotion at the limitations of human intelligence.'[4]

Whitehead, like J. B. Priestley, who is quoting him here, was a 'time haunted man', as are all of us who ask the question that stymied two of the greatest minds in history, and which kept more than a few who numbered between them awake at night. Time, for Whitehead, and for the rest of us, is one of the 'generalities' of our existence, those bottom-line imponderables that are 'incapable of analysis in terms of factors more far-reaching than themselves.'[5]

What this means is that any attempt to understand time will entail using terms that are just as imponderable as time itself. We cannot get behind or below time and in questioning time must accept as given the very time we are questioning.

This, of course, has not stopped us from us from trying to understand time, although those who do soon discover that T. C. Lethbridge was right when he said that 'time is one of those difficult words with many different meanings.'[6]

Time flows, doesn't it?

Take our basic notion of time as something that 'flows' from the past, to the present, and into the future; or, conversely, from the future, to the present, and into the past, depending on which direction we are facing. This seems unquestionable. 'Time, like an ever-rolling stream, bears all its sons away,' Isaac Watts wrote, hitting the appropriate note of loss. 'They fly forgotten, as a dream dies at the opening day,' Watts' hymn 'Our God Our Help' continues, providing an apt pairing, showing how time is subject to the same dissolution as are most of our dreams. If we don't like the stream analogy, we can think, as Andrew Marvell did, of 'Time's winged chariot hurrying near', another metaphor of motion aimed at capturing our experience of time. In one example, time is rushing away from us, from the present into the past. In the other, it is rushing towards us, the future hurtling at us; this is why Marvell doesn't want his coy mistress to 'waste time' on pointless preliminaries.

Before mechanical clocks achieved enough accuracy to make them popular – which happened sometime in the seventeenth century – the sundials they replaced were often inscribed with similar mottoes, announcing the irrevocable and hence depressing forward flow of time. Because whatever else it was headed towards, it was heading towards our death. 'As the long hours do pass away, so doth the life of man decay' and similar sayings give us an idea of how deeply ingrained is our perception of time as a 'sequence of events' occurring against a 'background of past, present, and future'.[7] As the physicist Danah Zohar writes, 'The whole rhythm of our conscious daily lives is lived out against the background of "passing time".'[8]

So much seems obvious. But if time is like a stream, where are its banks? We understand a river or a stream as a body of water flowing through a landscape. We can cross a river or stream from one bank to the other, as I do when I walk over Waterloo Bridge from the north to the south bank of the Thames. Yet, if time flows, it seems not to have any banks, in the sense of something alongside it that is not flowing, and it is just as impossible for us to imagine a river without banks as it is for us to imagine some kind of timeless, spaceless existence. Andrew Marvell's winged chariot is subject to the same question. Chariots roll on a road of some sort, and just as rivers have banks, roads are bordered by woods, or ditches, or pavements – something that is *not* the road and so helps to define it. What pavements flank Marvell's chariot?

In both examples we have an idea of a flow or movement of some kind, yet on closer inspection we see that the metaphor doesn't hold up. We *feel* that time flows, but we can't really picture it. When we try to with any seriousness, we see that if it flows, it is a different kind of flow from anything we are used to.

The direction of time

And the direction of the flow, from the past, into the present, and then on to the future, or is it the other way around? The past we know is gone; time's flow has carried it away. Precognition aside, the future has not happened yet, although time's chariot is racing towards it, it has yet to reach it. Where does that leave us? In the present. But no sooner do we try to grab hold of this present, than it is gone. The 'now' I can announce vanishes as soon as I announce it. It was the future a moment ago; now it is the past. The present, in which we always find ourselves – whatever time it is, it is the present – seems a kind of checkpoint marking a transition between two states of non-existence: a past that is 'gone', and a future that is 'not yet'.

But this present itself is strangely empty, merely an arbitrary point where the non-existent future instantaneously becomes the non-existent past. I look at the clock on my computer. It tells me it is 14:03. That is 'now', or at least it was. But it immediately gives way to another 'now.' And that new 'now' too cedes the throne of the present for the 'now' next in line, and so on.

Dreams may fade away on our waking, but with practice we can capture them sufficiently to write them down. No matter how hard I try, I cannot do the same with 'now'. Manufacturers of mobile phones sell us new models, offering more accurate photographic capabilities, assuring us that with their new camera we can 'capture' the moment like never before. But we never capture the moment, only an image of one that has inescapably become the past. We can look at that image and remember that moment, just as we can listen to a recording we may have made of our conversation with a friend that day. But we can never return to that moment itself – although there do seem to be some exceptions to this rule.

It's all relative

If we think that these reflections, disconcerting as they are, are really born of some misunderstanding, we should consider that no less an authority on time than Einstein gave them his imprimatur. 'People like us,' Einstein said, 'who believe in physics, know that the distinction between past, present and future, is only a stubbornly persistent illusion.'[9] The astronomer Fred Hoyle, a younger contemporary of Einstein who coined the term Big Bang – but never accepted the theory – agreed. 'The idea of time as a steady progression from past to future,' Hoyle wrote, 'is wrong.' It is, he said, 'a confidence trick.'[10] Yet even Einstein and Fred Hoyle set their alarm clocks to wake up in the morning.

We know that after Einstein, time became relative, and blended with space to form the curved 'space-time continuum' within which many science fiction films take place. This argued that if you were able to travel in a straight line indefinitely, you would eventually end up approaching yourself from behind; and if you travelled at the speed of light, time, for you, would stop. Actually, Einstein was preceded by the philosopher Gottfried Leibniz in this regard, who also argued that time was relative, but in a different way than Einstein. For Leibniz – who is counted as one of the early advocates of what we know as 'the unconscious' and was one of the first westerners to encounter the *I Ching*, the Chinese Book of Changes – 'instants, considered without the things, are nothing at all.'[11] Meaning that a time without something happening in it – without, as Whitehead would say, process of some kind – is unthinkable.

Leibniz

There is no time in itself, for Leibniz, nor space in itself, for that matter. Time is 'an order of successions … the successive order of things.'[12] Space is what separates two objects; without the objects, there is no space, just as without events there is no time. Leibniz saw, as we can ourselves, that it is impossible to imagine a time or space without *something* happening in them; just as in our attempts to imagine a timeless, spaceless existence, we draw a blank. Time and space, then, are relative to the processes taking place in them. For Leibniz, 'events' are 'more important than instants of time', which again seems an anticipation of Whitehead, who saw the universe not as a collection of things but as a series of events.[13]

Leibniz was not anticipating Einstein here, but was opposing his form of relativity to the understanding of time that Einstein's relativity dethroned: the absolute time of Leibniz's contemporary and sometimes rival, Isaac Newton.[14]

Let there be Newton

In *Philosophiae Naturalis Principia Mathematica*, published in 1687 and generally known as the *Principia*, Isaac Newton announced what for us is the 'common sense' understanding of time. 'Absolute, true, and mathematical time, of itself, and from its own nature, flows equably without relation to anything external,' Newton wrote.[15] And this flow remains constant and unaffected by any of the events happening within it, rather as a sea might remain unaffected by any ships that travel on it. 'All motions may be accelerated or retarded, but the flowing of absolute time is not

3: IT'S ABOUT TIME

liable to change.'[16] So an hour lasts just as long on the other side of the Milky Way as it does here, and there is nowhere in the universe where time flows faster or slower.

I should perhaps mention that an hour was not always of a uniform length. The idea of a day made up of twenty-four hours of equal length is in fact relatively new. The Egyptians, Babylonians, Chinese, and Japanese all had hours of unequal length; the Japanese did so up until the nineteenth century.[17] Nevertheless, one gets the sense that for Newton, time is not something in the universe, but that the universe is in time, and that time would remain, flowing equably as usual, whether there was a universe or not.

Entropy and absolute time

This is the relentless current of time from one second to the next that we have looked at. It is the time of which the philosopher Schopenhauer said that it 'never lets us so much as draw breath but pursues us all like a taskmaster with a whip.'[18] It is the time that led John Donne to remark that 'Man is a future creature', an assessment he shared with the philosopher Martin Heidegger, who saw the 'primary meaning' of human life as 'the future', 'that which comes towards one', Marvell's winged chariot. As George Steiner comments on Heidegger's notion about time and human being, what he calls *Dasein*, 'being-there', 'In seeking to be, *Dasein* is constantly ahead of itself and anticipatory.'[19]

For Heidegger, we 'project' ourselves into the future with plans and concerns, we *care* about what will happen, which we experience as anxiety, the *angst* that the Danish philosopher Kierkegaard believed we can escape only by reaching a rapprochement with the eternal.[20] Even science affirms this notion of an inexorably passing time, Einstein notwithstanding. The notion of entropy and the 'second law of thermodynamics' give us 'time's arrow', the irrevocable loss of order and organised energy in the universe that will eventually lead to its 'heat death' in a featureless cosmic soup, rather as a cup of coffee cools down to room temperature. This was an idea that depressed Victorian poets like Matthew Arnold and Alfred Tennyson with a sense of life's futility.[21] And although Newton's ideas about time, space, and gravity were superseded by Einstein's relativity, which made time and space relative to an observer, it is still Newton's time within which we 'live and move and have our being'. This was true even of Einstein, when he was not imaging what it would be like to travel on a photon at the speed of light, or predicting the existence of black holes, where, according to relativity theory, time stands still.

Abstract time

As we've seen, illusory or not, it is practically impossible for us to step out of this way of experiencing time. Aside from occasional moments when, for some reason, we find ourselves 'out of time' in a positive sense, we are for the most part carried along by this flow, riding in time's chariot, heading at an equable pace into the future, one tick of the clock at a time. What was epochal about Newton's announcement – which was practically dragged out of him; Newton was notoriously reluctant to publish his ideas – was not that he gave a proof that time flowed.[22] He didn't, and there were already many who had commented on time's inexorable passing. It was that abstract, absolute 'Newtonian time', shorn of any event happening within it, and oblivious to any consciousness experiencing it, passed from the world of mathematics into that of everyday life, and quickly came to dominate it. After Newton and the clockwork universe his *Principia* inaugurated, time was not the same. It had become disconnected from human events and been transformed into an impersonal, objective dimension.

What happened was that with Newton, time was disengaged from our experience of it. And until Newton's absolute time took over, that experience was rooted in the natural world. Well before the clock, the hourglass, the sundial, the clepsydra (an hourglass using water rather than sand) and other devices used by our ancestors to measure time, we were well aware of its passing through the events that marked the changes taking place in the natural world, the 'things' of Leibniz's relative time. The day was clearly marked by the sun's rising and setting. The phases of the moon could chart a month, while the seasons told of the earth's orbit around the sun, which gives us our year. For much of our existence, the earth itself was the only timepiece necessary.

The conquest of time

With Newton that changed. The clock, the most visible sign of Newton's victory, 'created a new reality', and was, according to the scientist Michael Shallis, author of *On Time*, 'as great an upheaval as was the invention of printing.'[23] When public clocks first appeared in the fifteenth century in European cities such as Prague, there was indeed a 'conquest of time', but not in the way that H.G. Wells meant in a book of that title. Time had been conquered by the clock. Time was no longer gauged by our felt experience of it, the appreciable change that communicates the different 'times' of the day, what the ancient Egyptians called *neters*, which is usually translated as 'god' but which might be better understood as 'vibration' or 'quality'.[24] Now time was something outside, quantified and mechanised, against which our lives were measured.

In the beginning

Animals live in a perpetual present. But when self-consciousness began to appear in humans, one of the first things we noticed was that the world we found ourselves in seemed to follow a certain routine. If dreams were important to our ancient ancestors, an awareness of time was even more so. Our concern with it goes back at least to the Neanderthals, circa 50,000 BC, who buried their dead, with ritual, decorations, and offerings of food, which suggests they believed in some sort of afterlife, that is, a future.[25] Cro-Magnon man, Neanderthal's successor, and our immediate ancestor, had an even more acute awareness of time. A bone dating from 35,000 years ago covered with a series of marks was recognised by Alexander Marshack, a research fellow at the Peabody Museum, as being a lunar calendar, charting the phases of the moon; it is perhaps the oldest 'timepiece' discovered so far.[26] It is now widely agreed that prehistoric megalithic sites, such as Stonehenge, were colossal calendars, marking important times of the year and keeping track of the heavens. Along with the rising and setting of the sun and the phases of the moon, our ancient ancestors soon became aware that the stars above them in the vast night sky moved in patterns, in stately ever-returning cycles. Man's first idea of time was cyclical.

A day, a month, a year are natural units of time, and are accurate enough for most of life, certainly before the modern age, which *is* modern precisely because of its different relationship to time. What made that relationship different is that time now became linear. The cycles of nature were superseded by the straight line, stretching endlessly into the distance. The hour, the minute, the second, and the finer temporal segments that our technological consciousness seems intent on differentiating, are not natural time units. They are, as it were, 'slices' of time, products of the absolute time that Newton bequeathed to us and which holds us in its thrall.

Martyrs of time

Some who have felt its grip struggled desperately to escape it. 'If you are not to be the martyred slaves of time, be drunk!' the nineteenth century French poet Baudelaire advises, an admonition that he, and many of his colleagues, often took in earnest.[27] A century after Baudelaire, another Frenchman, Albert Camus, was as aware of time's inexorable march as his predecessor. Where it led for him was 'the absurd'. 'Rising, tram, four hours in the office or factory, meal, tram, four hours of work, meal, sleep, and Monday, Tuesday, Wednesday, Thursday, Friday and Saturday, according to the same rhythm...' But, Camus tells us, 'one day the "why" arises' and with it 'the absurd', the sudden recognition that this routine, this life, is leading nowhere, and that perhaps all of human life is without meaning or purpose.[28]

This hurtling forward on the treadmill of time is not how humans have always experienced time, nor is it how all humans experience it today. Calendars existed long before clocks. It was important to know the right time of year to plant seed, but it was also important to have accurate dating for religious reasons, to mark the holy days and festivals that gave continuity to people's lives and offered them some participation with the divine. This time is what the historian of religion Mircea Eliade called 'sacred time', which is 'a primordial mythical time made present'.[29] Unlike profane or secular time, 'tick-tock time', sacred time is reversible 'by its very nature.' 'Sacred time,' Eliade tells us, 'is indefinitely recoverable, indefinitely repeatable.'[30] It is the time of the 'eternal mythical present that is periodically reintegrated by means of rites.'[31] It is a breakthrough by the *vertical* dimension of eternity, into the *horizontal* dimension of everyday life, the 'living time' about which Maurice Nicoll, a student of Ouspensky, wrote an important book.[32]

Sacred time

Sacred time participates in the 'eternal return', not in Ouspensky's sense of the repetition of our lives in eternity, but in the sense of re-enacting the primordial time of the gods, the time in the immemorial past when the acts commemorated and repeated in the sacred rites were first performed. That time has not left us, floating past in the stream, nor has it been carried away by a winged chariot. It is ever-present, existing apart from this unceasing flow. The religious rites and festivals that facilitate the eternal return 'have no part in the temporal duration that precedes and follows them.'[33] They are 'out of time' and constitute a patch of eternity made present in the profane world. It is this kind of time that the Australian aborigine enters when he leaves the world of everyday time and journeys along the 'song lines' of his ancestors into the Dreamtime, the mythic dimension in which the contours of the land were laid out, not in an inaccessible past, but in a time that runs parallel with that of the clock.

Our holidays – 'holy days' – are supposed to serve the same function as these excursions into an ever-present mythical dimension of life. They are supposed to punch a hole in the ordinary succession of events, to cause us to stop and remember that we do not live by bread alone. Yet it is not difficult to see that in our utilitarian society, run by the clock, 'holy days' have ceded practically all of their power to the practical exigences of the secular world, and an ungenerous observer might remark that for us, it seems that Jesus was born and died so that we could enjoy three-day weekends.

Although it is not impossible for people today to enter the sacred time that the rites of their religion can open to them, I think we must admit that such participation in the eternal is rare for us moderns, harried as we are by Newton's equable flow. Yet we do not have to look at ancient peoples to get some sense of what experiencing such a time would be like. In his book *The Dance of Life* the anthropologist Edward T. Hall describes the radically different sense of time that he discovered while living and working with different tribes of Native Americans, the Hopi, Navaho, Pueblo and Quiché. What Hall discovered was that the time sense of these peoples is so different from our western one – what Hall calls the AE, American-European, time sense – that it amounts to a completely different kind of time.

Polychronic time

One thing that Hall discovered is that the Hopi have no word for time, and that their verbs have no tenses. Like the participants in the rituals that evoke Eliade's 'sacred time', the Hopi live in an 'eternal present', 'indifferent to western science, technology and philosophy.'[34] We westerners, Newton's 'martyred slaves', live in what Hall calls 'monochronic' time, an apt name given that for many of us such time is monotonous. The Hopi and other Native Americans Hall came to know, live in a 'polychronic' time, a time of multiple character, in which they feel and are sensitive to the same differences in time that ancient Egyptians felt in their *neters*.

Hall found that for the Hopi, each day had its own character, its own shape, and was not 'just another day', as it is for us. This was something that Ouspensky knew as well. In his early work on the mysteries of time and space, *Tertium Organum*, Ouspensky wrote that 'a peculiar mood of their own is felt in certain *days*.' 'There are days,' Ouspensky tells us, 'full of strange mysticism, days which have their own individual and unique consciousness, their own emotions, their own thoughts.' To us this may sound absurd, but the Hopi would agree. They would also find nothing absurd in Ouspensky's suggestion that 'One may almost talk with such days.'[35] The Hopi would say that this is exactly what they do. 'Time has a quality about it...that is instinctively grasped,' Michael Shallis writes, and unlike us, the Hopi it seems are very good at grasping it.

Hall points out that 'religion is the central core of Hopi life' and that it is because of this that they live in their 'eternal present'. As the ancient Egyptians were, another people for whom religion was central, the Hopi are aware of meanings and significances in time of which we Westerners are for the most part ignorant. One suggestion about the motive ancient and prehistoric peoples had for keeping track of what was happening in the heavens – and for building immense structures such as Stonehenge – is that they were somehow attuned to energies and forces in nature to which we have become insensitive. These energies or 'qualities' ebbed and flowed in a rhythm that could be recognised and the stone circles and other prehistoric monuments that dot Europe may have been erected in order to mark the times of the year when they were most powerful.

With the rise of our modern way of engaging with the world, our ability to feel these energies became dulled or forgotten and was eventually denied, but the Hopi it seemed were still able to detect their presence. One reason for this is the different way they experience time.

Bergson and duration

Westerners are always in a hurry. When C. G. Jung met the Pueblo chief Ochwiay Biano during a trip to Taos, New Mexico, the chief told him that the whites he met were always dissatisfied, and that this was the reason for their cruelty. Jung encountered a similar feeling in Africa; in both places what was involved, Jung believed, was a different sense of time. In these places, where time had not yet been 'conquered' by the clock, there was a 'dream of a static, age-old existence', the perpetual now of myth and the soul. Yet this dream would not be allowed to continue indefinitely. Jung found himself thinking of his pocket watch – he came, of course, from Switzerland, a land of cuckoo clocks – and he wondered how long it would be before the 'god of time' and his demon, progress, would 'chop into bits and pieces' the 'duration' he felt there, which, he said, was 'the closest thing to eternity'.[36]

'Duration' – inner, experienced time as opposed to the Newtonian brand – is most closely associated with the philosopher Henri Bergson, whose work Jung knew. In books like *Time and Free Will*, Bergson put the time question firmly back on the philosophical map, even having a famous debate with Einstein about it.[37] Although he had 'relativised' Newtonian time, Einstein argued that only physical time, that measured by scientific instruments – increasingly fine-tuned clocks – was 'real' time. Bergson argued the opposite: that this scientific time was only possible because we had an intuitive sense of time, what he called 'duration', which is the kind of 'felt' time Hall discovered among the Hopi. In duration time is not experienced as a succession of seconds, nor as broken up into past, present, and future. It is marked by what we can call 'changes of state'.

In the end it was all relative. For the popular mind Bergson's duration lost out and the 'real' time Einstein defended won, with results that Einstein himself would not necessarily have appreciated. As Michael Foley writes in his little book on Bergson, the popularity of modern movements such as the Quantified Self, 'which promotes, as a key to well-being, a self constantly measured and monitored', suggests that the 'tick tick tock of the stately clock' – with apologies to Cole Porter – has come to dominate us increasingly despite Bergson's arguments.[38]

A change in time

Yet the interesting thing that Hall discovered is that after spending some time with the Hopi, his own time sense began to change. Hall writes of going on a long ride to help a friend bring his horses from New Mexico to Arizona. They could travel only fifteen miles a day, so that the horses did not tire, and the long hours spent at this leisurely pace began to affect how Hall perceived things. He 'watched the same mountain from different angles during three days,' and it seemed to 'slowly rotate' as they passed it. 'Experience of this sort,' Hall remarked, 'gives one a very different feeling than speeding by on a paved highway in one or two hours. The horse, the country, the weather set the pace.' On another longer ride, one of a few hundred miles, Hall found that 'it took a minimum of three days to adjust to the tempo and the more leisurely rhythm of the horse's walking gait.'[39]

Hall was moving out of the time sense of a modern Westerner, regulated by the clock, into something more like how the Hopi experience time.

It is interesting that Hall says it took a minimum of three days to adjust to the slower tempo. I have not ridden a horse for hundreds of miles, but I do know that on different occasions I have experienced something similar to what Hall is describing. It happened once when I was living in Los Angeles, and had decided to drive north to the sequoias to rent a cabin for a week. The sequoias are gigantic trees that live for thousands of years; they have, we could say, a time of their own. I was looking forward to getting away from the city and losing myself in these ancient forests. But what I found was that I had brought the city with me. For the first few days I found it almost impossible to relax. Even with the best intentions I was anxious and on edge, unable to adjust to the cabin, constantly looking for the things that usually occupied me and whose absence irritated me. It wasn't until the fourth day that the irritation lifted and I found that I could 'be' there without complaint. I then spent the remaining days wandering among the colossal redwoods, perfectly content. I had slipped from tick-tock time to duration.

As I write this, I am reminded of another occasion. I was in Cornwall, years ago, walking the coastal path. It was summer, in the middle of a heat wave, and at one point I found myself sitting near crystal-clear water, peering at pebbles along the shore, glistening in the sunlight. To this day I don't know how long I sat, looking at those pebbles, but it seemed like hours. If I try hard enough I can conjure the *feeling* of being perfectly contented, perfectly quiet, and perfectly happy doing nothing but looking at them. Again, duration.

Slow down

In my case, and even more in that of Hall, what had happened is that time had *slowed down*. Or rather, we had. And one result of this slower pace was that we saw things that we ordinarily wouldn't have. This was something I noticed on a more recent occasion, when knee surgery had me on crutches for weeks, and I was forced to move at a much slower pace. Hobbling back to my flat one day, I found myself fascinated by the reflection of sunlight on the flecks of mica in the pavement. I walk on pavements all the time. But it wasn't until the crutches forced me to slow my usual pace, and to *concentrate* on each step, that I *saw* the glitter and sparkle beneath my feet, which are always there but of which I am usually unaware. Even the cracks in the pavement were interesting and seemed suggestive of something. As Hall had, by slowing down I *saw* more.[40]

I don't say I slowed my 'normal' pace, because our usual pace isn't normal, but is a product of Newtonian time. It may flow equably at the same speed at all times. But the creatures that are most aware of it – ourselves – find themselves pushed to increasing acceleration by it.

But what does it mean to say that time 'slowed down?'

Before we try to answer that question, try this experiment.[41] Take an old-fashioned watch – not a digital one – and look at it. You can see the second hand move, but not the minute hand and certainly not the hour hand. Now, put the watch down and relax. Close your eyes and imagine some beautiful scene, a tropical beach or sunset in the desert or some other pleasant daydream. Try to visualise the scene as vividly as possible: hear the waves, feel the sun, smell the salty air. Now open your eyes and casually look at the watch, detached, as if your gaze just fell upon it by chance. If you have relaxed deeply enough and if you have visualised your inner scenery vividly enough, you should notice something. You should see the second hand *stop* and hover in place briefly, where before it seamlessly glided past. You might think that your watch has stopped. It hasn't. It has not slowed down; *you* have. You should also be able to see the minute hand *move*, not in jerks like the second hand, but slowly, glacially we might say. There is no reason that with enough practice you should not also be able to see the hour hand move. We *know* the minute and hour hands move, but we usually don't see this. But the relaxation and visualisation, if done seriously, should slow our usual forward motion enough for us to see what we usually are blind to, just as my knee surgery, which had forced me to slow down and concentrate on each step, allowed me to notice things I usually ignore.

A variation on this exercise is to look at clouds on the horizon. It requires fine weather and a calm day, without a breeze, and it helps if you can get to a hilltop. Pick a cloud that, at a glance, appears motionless. Then relax and look at it as you would if you were expecting something and watching out for it. 'Relaxed alertness' is what is called for, a 'calm expectancy'. You should soon notice that you can see the cloud move, again, slowly, like the minute hand.

What these experiments tell us is that what we take to be our normal or natural sense of time is not normal or natural at all. With a little effort we can change how we experience time. And in changing our experience of time, we change our experience of the world.

How do we do that?

Our two brains

One answer to that question is that what happens in these moments is that we move from our usual sense of time, which is associated with our left brain, and into a broader, deeper time sense, which is associated with our right brain. We can say that Newtonian time is left-brain time; Hopi time is right-brain time.

The fact that we have two brains, or that our brain is divided into two cerebral hemispheres, connected by a knot of fibres known as the corpus callosum, or commissure, has been known for millennia. Greek physicians of the third century BC were aware of it, and they were also aware that the two brains worked differently. The right brain, they saw, was geared towards 'perception', while the left seemed to be associated with 'understanding', a division of labour that has held up with time. In more recent years a great deal of popular literature about our two brains has led to what for some has become a cliché, with the right brain being an 'artist', and the left a 'scientist'. Subsequent research has qualified this assessment somewhat, but for our purposes here it stands. The left brain deals with logic, language, and sequential thought; the right is geared towards recognising patterns, and is more at home with images and intuitions. This handy distinction is supported by clinical evidence. As Colin Wilson writes in his book about the double brain, *Frankenstein's Castle*, 'People who had damage to the right cerebral hemisphere were unable to recognise simple patterns or enjoy music, but they could still speak normally. People with left brain damage were able to recognise patterns, but their speech was impaired.'[42]

Exactly *why* we have two brains remains a mystery, but research in split-brain psychology – when the commissure connecting the hemispheres is severed – arrived at a remarkable conclusion: that 'we' are literally two people. The person I know as 'I', my verbal ego, lives in the left brain. Next door in the right cerebral hemisphere lives a stranger, who doesn't use language, but who speaks in symbols, images, metaphors, moods, even physical sensations.

Who is this stranger? That is a good question. Whoever it is, he or she seems to be responsible for our moments out of time and for the strange world we inhabit in our dreams, as well as for odd experiences such as synchronicities and other paranormal phenomena. In *Faust* Goethe lamented that 'Two souls, alas, live in my breast.' He may have got his anatomy wrong, but the central point is sound. We literally are two people, only one of whom we know relatively well.

A cerebral coup

In recent years, the split-brain discussion, which had been abandoned by 'serious' neuroscientists because of the interest shown in it by New Age enthusiasts and pop psychologists – both camps nevertheless contributing important insights – has been rebooted by Iain McGilchrist in his book *The Master and His Emissary*.[43] Briefly, McGilchrist argued that although subsequent research found that the neat separation of function that was originally thought to characterise our two brains was more complex than initially believed, the distinction between the brains remained and was crucial. The difference between them was not so much in *what* they did, but in *how* they did it. Our two brains do things very differently. McGilchrist's other important point is that the left brain, long believed to be dominant, leaving the right brain as a sometimes helpful sidekick, was really a usurper. It is the 'emissary' of the title of McGilchrist's book, while the right brain, which is older, is the rightful 'master'. Somewhere along the line the left effected a *coup d'état*. This was something Colin Wilson had also recognised. Writing of the relationship between the two brains, Wilson remarked that 'the being who calls himself 'I' is a usurper. It is his brother, who lives next door, who is the rightful heir to the throne.'[44]

The forest and the trees

What is the difference between the two brains? As McGilchrist tells us, the right brain perceives the world holistically, as a total, global picture. It is geared towards engaging with living things, and is interested in overall patterns, meanings and relations. It is open to the new, the unfamiliar, with what the medieval Rhineland mystic Meister Eckhart called *Istigkeit*, the 'is-ness' of things, the sheer awareness of their unique reality, what in Zen Buddhism is called *satori*. It is also geared towards what the philosopher Michael Polanyi called 'the tacit dimension', the implicit, intuitive meanings of which, as Polanyi says, 'we can know more than we can tell.'[45]

When St Augustine tells us that if we do not ask the question 'What is time?' he knows the answer, it is his right brain that knows. But it does not speak, at least not in language. That is why he can't *tell* us the answer to the question, which his left brain, which does use language, asks and which leads, as we've seen, to paradox and confusion. The right brain communicates in metaphor, images, symbols, even physical sensation, which can convey the *implicit* meanings that time and the other imponderables mentioned earlier – Whitehead's 'generalities' – evoke. In this the right brain seems rather like our dreams, which also use metaphor and symbols and which often seem to have their own peculiar way with time. These implicit meanings cannot be made explicit; they cannot be nailed down in a formula, they are the 'more' that we know but cannot 'tell'. This is why J. B. Priestley remarks that 'Truth can be obtained only at the expense of precision.' The kinds of truths that are truly important are those that we can never speak of explicitly. We can show them or hint at them, as a painting or a poem or symphony can. But we can never capture them in a neat syllogism.

The left brain's job is to unpack the vague, general, total picture presented by the right brain, to give it detail and clarity. If the right brain sees the forest, the left sees the trees, and the individual leaves on the tree, and the individual veins in the leaf, and so on down to its very cells. The right brain presents the 'big picture'; it gives us a panoramic view. The left breaks this down into its parts; it shows us the world through a microscope. The left's job is to help us manoeuvre through the world, to make it manageable, to control it. And the best way to control the world is to take it apart and make it familiar. While the right brain is open to 'newness' and seeks it out, the left aims to reduce everything to what it already knows.

Eliminate the inessential

We can see that each has its drawbacks. The right has a strong but imprecise sense of meaning and connection to the whole; it 'knows', but can't tell you, except through hints and suggestion.

The left has precision, but its pinpoint accuracy is achieved through loss of connection to the whole, and hence any sense of meaning. I should also point out that if the left has a manic optimism and forward drive, the right tends to depression. So, whether we like it or not, we are all manic-depressives, at least insofar as our cerebral hemispheres are concerned.

The left brain's penchant to break up the whole presented by the right, in order to manipulate reality, can be seen in its approach to time. Most research suggests that the left brain is obsessed with time – 'stupidly' so – while the right brain is cheerfully oblivious of it.[46] This obsession with time makes for punctuality, but it also results in some problems. Bergson recognised these long ago, although he didn't have split-brain psychology to support his intuitions. Bergson argued that the brain is an 'organ for survival', and that its function was essentially eliminative. Its job is to filter out stimuli and information that isn't essential to dealing with the necessities of existence, and stop it from reaching consciousness; this means 90% of reality. (We know now that it is the left brain that performs this function.) What this means is that our intellect – or left-brain consciousness – is useless when we try to grasp the nature of reality.

The mind is the slayer of the real

Say I draw a line on a sheet of paper. My left brain tells me that it is made up of innumerable points, each one occupying its part of space. But I *see* a continuous line, not a series of points, and when I drew the line I did it in a continuous motion. And even if I did see each individual point, I could divide any point even further, ad infinitum. The same is true for time. I can divide an hour into minutes, minutes into seconds, seconds in microseconds, and these into even shorter segments: milliseconds, nanoseconds, etc. My intellect tells me there is no reason why I can't continue dividing time into shorter and shorter units, never reaching any kind of endpoint. This is the thinking behind Zeno's paradox, in which the arrow never reaches the target, because there is always a shorter bit of time and space remaining *before* the arrow hits. The same goes for Achilles who never passes the tortoise, and so loses the race. Yet we *know* that the arrow reaches the target and that any runner can overtake a tortoise. Zeno devised these paradoxes in support of Parmenides' argument that time and change is an illusion. But what they really tell us is that our left brain is hopeless when it comes to reality. 'The mind,' Madame Blavatsky said, 'is the slayer of the real,' an insight with which Bergson would have agreed. Time, as Colin Wilson says, 'is an invention of the left brain.'[47] It is a very helpful invention, which aids us in our mastery of the world. But if we try to understand reality through it, it leads to a muddle.

Right brain stasis

Yet, although a right brain approach to time frees us from our temporal taskmaster, it presents problems of its own. Jung put his finger on this when he said that what he found among the Indians of New Mexico and in his encounters in North Africa was a 'dream of a static, age-old existence'. If the left brain has a relentless forward drive and is always in a hurry to get things done, the right is happy to lollygag with its hands in its pockets, going nowhere in particular. This may seem like a holiday when we are stuck in the rat race, but the problem here soon becomes clear. Right-brain consciousness is static; if the left brain is obsessed with time, the right ignores it. It can 'be here, now' without a care for tomorrow. That is why I could sit, gazing at pebbles for hours, without worrying about the time. But a cow is here now. And although, like Walt Whitman, we may envy cows their placid being, we cannot ask them the secret of their happiness because, as Nietzsche points out, they would forget their answer before they could tell us.[48]

If the right brain is the master, it created the left as its emissary *for a reason*. What reason? The short answer is to evolve, to strive, to face challenges and overcome them, to develop our powers that require purpose in order to actualise. Left-brain consciousness forces us to struggle for freedom and to 'become ourselves', to individuate, as Jung would say. The limitations of left-brain consciousness turned the broad river of the right into a powerful jet. That narrowing, that focus, is what has enabled us to master the world, but it has also alienated us from it. Many of us find our left-brain consciousness a burden and dream of giving it up and 'getting back to nature'. But nature it seems cast us out of its warm embrace and into the cold light of left-brain consciousness on purpose. This was no 'fall', as many romantics, yearning to return to some unself-conscious innocent state, claim. It was more of a leap.

Aldous Huxley's dishes

One method of achieving this desired 'return to the garden' is through the use of psychedelic drugs and other mind-altering substances that we know affect our sense of time. Baudelaire advised being drunk for precisely that reason; a glass of wine relieves the pressure of the clock. Other substances achieve an even greater release. While under the influence of mescaline, Aldous Huxley was asked what he thought about time. 'There seems to be plenty of it,' Huxley replied, expressing the 'indifference to time' he felt under the drug. After taking mescaline, Huxley experienced 'an indefinite duration or alternatively a perpetual present.'[49] Mention of 'duration' reminds us that Huxley accounted for the effect of the drug by referring to Bergson's notion of the brain as an eliminative organ, allowing only so much reality into consciousness as 'will help us to stay alive on the surface of this particular planet.'[50] Mescaline had turned the brain's filter off, and all the reality that had been edited out of Huxley's consciousness came rushing in, with the effect that he felt he was 'seeing what Adam had seen on the morning of his creation – the miracle ... of naked creation.'

But Huxley saw that this miracle came at a price. 'The will [under mescaline] suffers a profound change for the worse. The mescaline taker sees no reason for doing anything in particular.' Dishes in Huxley's sink were too beautiful to wash – as many less sober psychonauts discovered – and Huxley came to the conclusion that if everyone took mescaline there would be no war, but there would be no civilisation either, because no one would be bothered to create it. And although there certainly are problems with civilisation – the obsessive awareness of time's relentless pressure being one of them – Huxley knows that it is certainly worth the trouble of creating it.

An earlier form of consciousness

Mescaline and other drugs return us to an earlier form of consciousness, in which the left brain's obsessiveness is muted, our experience of time is altered, and in which metaphor, symbol, and image, not language, are the means of communication. If Arthur Koestler, who also experimented with psychedelic drugs, is correct in saying that in the dream, 'rational controls are relaxed and … the mind seems to *regress* from disciplined thinking to less specialised, more fluid ways of mentation' – a 'retreat from articulate verbal thinking to vague, visual imagery', which sounds very much like a switch from left- to right-brain consciousness – we can reach some tentative conclusions. Drugs like mescaline and less potent substances like alcohol, inhibit the obsessive left brain and shift us into a more right-brain awareness. This right-brain consciousness is older, and it created its partner, the left, because its work of filtering and organising our experience was necessary. Without it, we wouldn't have left our animal past. We may not like leaving it, but there's no going back.

In *Lost Knowledge of the Imagination* I point out that there is good reason to believe that our ancient ancestors enjoyed a consciousness that was image-based and intuitive and not logical and language-oriented as ours is. Yet this more image-oriented, dream-like, right-brain consciousness was superseded by our language-based, daylit, rational, left-brain consciousness. Exactly when this shift began is unknown – the psychologist Julian Jaynes put it at around 1250 BC – but it was certainly on its way by a time known as the Axial Age, circa 800 to 400 BC.[51] It was during this time that philosophy began in Greece, and with it the tradition of rational inquiry that has come to characterise the west. Newton, of course, didn't introduce the left brain – although he had an exceedingly good one – but by the time of the *Principia*, its dominance was assured. As I argue in *The Secret*

Teachers of the Western World, by that time, there seemed to have been an all-out attack on the Hermetic tradition, which I believe can be seen as a repository of right-brain knowledge and ways of knowing. We see the effect of this today, with a reductive, materialist scientism declaring that dreams, often a source of Hermetic wisdom, are meaningless.

The servant does his job too well

The left brain, however, is not a villain. As we've seen, the right brain, the master, needs him. We all do. I could not have written this book without him, and you would not be reading it, nor would we enjoy the standard of living we do, which kings of old could not have dreamed of. Without doubt, the left brain has brought us many good things. But, as we've seen, there is a problem. The two brains are supposed to work together, and they each have a stake in the other, just as in the yin-yang symbol, an ancient intuitive image of our polarity, each side has a bit of the other.[52]

Yin-yang.

What has happened is that the emissary has usurped power, which is another way of saying that it *does its job too well*. It is like the demi-urge in the ancient Gnostic myth, who was employed by the true God in order to create the world, but who forgets his brief and comes to believe that *he* is the true God and acts accordingly. Christians call this Original Sin, but it is not a sin, merely a mistake, a kind of delusion. It is an evolutionary fault, that is, a kind of habit, and habits can be changed. It has certainly landed us in some difficulties, but ultimately the prognosis is good, because we can teach our left brain – that means ourselves – to rectify the imbalance. I suspect the ever-patient right brain is waiting for us to do just that.

Chapter Four
Looking Ahead

Impossibility of precognition. Psychometry. Temporal photographs.
Before the Dawn. *Louisa Rhine. F. W. H. Myers at a loss for words.*
Stan Gooch, dreams and the paranormal. Danah Zohar. Koestler.
The trouble with time travel. Evidence for precognition. Society for
Psychical Research. Quantum weirdness. Dean Radin. Time Loops.
Lord Kilbracken picks a winner. Wilbur Wright does too. David
Mandell's dream paintings. Aberfan. Dickens and Miss Naiper.
Schopenhauer's ink spots. Lincoln. Sarajevo. Dunne. Intrusions.
The spirits speak. Serialism. Priestley and Ouspensky. The mother,
the creek, and the baby. Your life is your time. Eternal recurrence.
Reincarnating in the past. Lethbridge and his pendulum.

On the face of it, precognition should be impossible. It simply
shouldn't happen. We could account for telepathy, clairvoyance,
even psychokinesis – the ability to move objects simply by the
mind – and other paranormal phenomena in terms of some kind of
mental force that science hasn't yet discovered. Some kind of 'mind
ray' could transmit thoughts from one mind to another in the way
that radio waves transmit information from a sender to a receiver.
In fact, years ago, the novelist Upton Sinclair wrote an account
of his experiments with telepathy entitled *Mental Radio*; Einstein
wrote a preface to it. Something along the same lines could explain
'remote viewing' and even psychokinesis, with 'thought energy'
of some sort applying pressure to physical objects. Attempts to
account for paranormal phenomena in this way have so far failed,
not because of some conceptual impossibility, but because to
have knowledge of some event that hasn't happened yet seems to
contradict everything we know, or think we know, about reality.

Even the idea of investigating the past through what is known
as psychometry, the ability to gather information about the history

of some object simply by holding it, which strikes us as incredible, does not transgress against logic in the way that precognition does. Unlike the future, the past has happened, and just as we have a variety of ways of 'capturing the moment' – through photographs or recordings – it could be discovered that objects too, in some way, carry a record of their past. The mathematician E. T. Bell wrote science fiction stories under the name John Taine. In *Before the Dawn*, a character invents a 'light decoder', a device for analysing the different 'photographs' that have accumulated as 'double exposures' made by sunlight falling on the surfaces of different objects. The principle is the same as ordinary photography, so conceivably such a device could be invented.[1] But no device can analyse photographs that have yet to be taken.

In our world of the Internet, artificial intelligence, quantum physics, and science fiction films that have accustomed us to concepts like parallel worlds, alternate realities, and time travel, it is easy to lose sight of just how odd precognition is. But if we have become inured to it, that is a reflection on the dubious human ability to get used to practically anything, not on precognition. That it should happen should stop us in our tracks, and if we are lucky, it does. As Colin Wilson writes, 'if we are willing to admit that ... on one single occasion, Dunne ... actually dreamed of something that had not yet taken place – then we have admitted the possibility that the common sense view of time is as crude and simplistic as the flat-earth theory.'[2] That many of us do not recognise the problems in our view of time, or if we do, shrug them off as unimportant, suggests that Columbus or not, many of us still live on a flat-earth.

Materialism overturned

But not everyone. Others who have researched and experienced precognition have been truly stunned by its implications. Louisa Rhine, wife and colleague of J. B. Rhine, who put parapsychology on the academic map at his Parapsychology Laboratory at Duke University, North Carolina, in the 1930s, was one of them. As she writes, 'Perhaps no type of ESP experience seems more incredible ...

than [the] type which involved the future.' Yet what is odd about this is that precognitive experiences 'are reported more frequently than any other type of ESP.'[3] (ESP, of course, meaning extra-sensory perception.) F. W. H. Myers, one of the founders of the Society for Psychical Research, and author of *Human Personality and Its Survival of Bodily Death*, was almost at a loss for words when it came to precognitive dreams. For him precognition involved 'a category of phenomena which at present I can make no attempt to explain.' Myers was stymied by 'its definiteness, its purposelessness, its isolated unintelligibility.' Such dreams, he admitted, were 'more interesting' than others, but they were also 'more perplexing.'[4] Speaking of precognition, Stan Gooch, a psychologist and medium with first-hand experience of the paranormal, writes that 'All the armour and edifice of science is of no avail against it. With this one stroke' – the precognitive event – 'the materialistic universe' – including Newtonian and Einsteinian time – 'as a *total* explanation of events, is in ruins.'[5] I should point out that for Gooch, 'Dreams ... make the single greatest contribution ... to our understanding of paranormal phenomena', and precognitive dreams make the greatest contribution of all.

If we think that these statements, coming from confessed psychic explorers, may be biased, how does this sound, coming from someone in the world of physics? 'Solid proof that some people do indeed have foreknowledge of events in the future,' writes Danah Zohar, 'would challenge the most fundamental tenets of both common sense and classical physics.'[6] Arthur Koestler, who moved from politics, to science, to parapsychology, sums it up with typical restraint: 'The third category of ESP phenomena, precognition, including premonitory dreams, seems to have a higher degree of incredibility than telepathy and clairvoyance.'[7]

There's no 'there' there

Why should precognition be harder to believe than telepathy, clairvoyance, or psychokinesis? Because it seems to depend on a kind of time travel, specifically into the future. For our common sense view of things, the future does not yet exist, so technically, as Gertrude Stein once said of her hometown of Oakland, California, 'there's no "there" there.' If the future does not yet exist, there would be no time or place to travel to. If we think of time as a railway track, the track behind us stretches into the hazy distance. The track ahead of us hasn't been laid out. Or rather, it is being built, one second at a time.

Yet every so often we hear reports from scientists who tell us that time travel *is* possible, at least theoretically. Einstein's relativity proposes the possibility of 'wormholes' in space-time that would allow someone from 'here-now' to travel to 'there-then', whether that means the future or the past in this galaxy or another.[8] The physicist Richard Feynman even proposed that one elementary particle, the positron – a kind of negative electron – is really an electron moving backwards in time, rather like the character in Algernon Blackwood's story, 'The Man Who Lived Backwards'.[9] Some scientists go further and propose an entire other universe moving backwards in time.[10] I won't go over the various ways physicists suggest for travelling in time, which fill the pages of pop science books, lend plausibility to scores of sci-fi films, and inform contemporary television shows like *Dark* and *Devs*. But it strikes me that there are at least two logical problems with time travel in the traditional sense, starting with H. G. Wells and *The Time Machine*.

I should point out that Wells was a friend of Dunne. In *The Shape of Things To Come*, one of Wells' 'future histories', the main character, Philip Raven, has read *An Experiment with Time* and has learned how to extend his own future dreams so that they reach well beyond the next day, week, or month, into the very far future.[11] In fact, we could say that the time trip the anonymous hero of Wells' early work takes and that of the later Philip Raven are rather different, in a way that is essential to this book.

The trouble with time travel

In *The Time Machine* Wells' hero, known simply as the Time Traveller, invents a device that can take him into the future or the past. As many did at the time, Wells saw time as a 'fourth dimension', added to our usual three. Just as we can travel freely in space, the time machine allows the Time Traveller to travel in time.

Let's think about this. If the objection to travelling into the future is that it does not yet exist, surely this shouldn't be a problem if we travel into the past, which at least has happened. But just to be safe, say you travel back only one day. You go back to yesterday and tell yourself, whom you meet there, that you shouldn't worry, time travel is safe. Then, to have company on your trip, you take your past self along, and you go back a week. There the two of you – literally – meet yourself from a week ago. So that he won't feel left out, you take him along for another trip. While there you encounter yourself yet again. Conceivably you could travel back into the past innumerable times picking up earlier selves. Forget the nagging fact that in each of these past times, you did *not* meet your future self and were not thrilled to take a ride in the time machine. So technically, the past you go back to *isn't* the past that you left behind, because you didn't meet yourself in that past. This reflection raises the problem of the 'time loops' that bedevil attempts to take time travel literally, with some scenarios having an individual become the cause of his own arising, 'bootstrapping' himself into existence. But even without this problem, the absurdity is that on your forays into the past, you could soon collect an infinite number of selves, far too many to fit for a return trip.

Another objection arises from the idea of time travel itself. If time travel was *ever* possible, then by definition it should *always* have been possible. Think about it. Say someone invents a time machine in the year 2120. They use it to go back to 2020. Is time travel possible in 2020, speculation by physicists notwithstanding? If so, it hasn't made the news. Say they go back to 1920, 1820, 1720, and so on. If they do, then time travel was possible in those times, just as it would be in 10,000 BC. If you are travelling in time, then time presents no problem. It doesn't matter 'when' a time machine is invented. Once it is, 'when' no longer matters. So if someone invents a time machine, time travel is then possible *at any and all times*, just as 'space travel' is possible in any and all spaces. As far as we know, there are no 'spaces' we can't travel through, contrary to unhelpful folk who, when asked directions, reply 'You can't get there from here.' There is no 'there' that you cannot get to from any 'here'. The same is not true of time.

Unless it has been kept an incredibly secure secret, time travel in the traditional sense is not possible. To say 'still not possible' or 'not yet possible' suggests that you miss the point.[12]

Yet in *The Shape of Things to Come*, Philip Raven does travel, not into the past but into the future, and not with a machine but in his dreams, that is, *with his mind*. We will return to this important difference.

Science says it's for real

At a seminar held at the Omega Studios in Rhinebeck, New York, in October 2019, I mentioned my doubts about time travel to one of the other speakers, Dean Radin, one of the most important scientists working on the paranormal today. I can't remember his exact words, but his reply was something like: 'This might be true in an Euclidean universe, but not in a non-Euclidean one.' As I had struggled through Euclid in school, I wasn't equipped to challenge the point, although I knew of non-Euclidean geometry and 'higher space' from reading Ouspensky and other writers on the subject. But it still struck me that logic was on my side. It still does. I am open to the idea of *seeing* the future, and also the past. I just don't think that I can hop into my time machine and go there, in some literal, physical sense.

At the seminar, Radin spoke about the wealth of scientific proof for paranormal phenomena such as precognition which, in the words of the statistician Jessica Utts, 'would be widely accepted if it pertained to something more mundane.'[13] The theme of the seminar was 'Real Magic', the title of a book by Radin. What is 'real magic?' Basically it is the paranormal which, Radin argues, has by now acquired enough statistical evidence to prove its reality. He makes the point that much of what we consider magic and 'the occult' are manifestations of paranormal abilities. This was something I already believed, so I was happy to accept his invitation to speak at the seminar.

Society for Psychical Research

As I had been convinced of the reality of the paranormal for more than forty years, I was happy to see that some scientists are finally coming round to accept it. I knew that others had accepted it in the past. In fact, in the early days of psychical research, some of the most prestigious names in science and philosophy believed the paranormal more than worthy of exploration. A short list of members of the Society for Psychical Research would include William Barrett (physicist), Lord Rayleigh (physicist), Arthur Balfour (philosopher and British Prime Minister), Eleanor Sidgwick (mathematician), Charles Richet (physiologist and Nobel Prize winner), William Crookes (chemist and inventor of the Crookes tube), Oliver Lodge (physicist), Henri Bergson (philosopher and Nobel Prize winner), to name just a few. William James, the great American philosopher and psychologist, was also deeply involved and started an American branch of the society.

It is eye-opening to go back a century and see a Nobel Prize winner like Bergson openly discussing his interest in parapsychology and mysticism. The difference between then and now is apparent in an oft-quoted statement from the astrophysicist Steven Weinberg, himself a Nobel Prize winner. Weinberg closes *The First Three Minutes*, his account of the Big Bang, by saying 'the more the universe seems comprehensible, the more it seems pointless.'[14] Bergson closes his last major work, *The Two Sources of Morality and Religion*, with the reflection that the 'essential function of the universe' is that it is 'a machine for the making of gods.'[15] The disparity between the two views couldn't be more clear.

So for scientists and other academics of high standing to be interested in the paranormal and to even affirm its existence, welcome as it is, is not new. Yet the relationship between science and the paranormal has been rocky at best. After the initial honeymoon, an estrangement set in, which was ameliorated somewhat when J. B. Rhine opened his Parapsychology Lab. Yet as Arthur Koestler pointed out, one result of this was that the sensational, mysterious, mystical world of the paranormal soon metamorphosed into one of the most dry as dust disciplines in the academy. In order to prove itself worthy, parapsychology had to be more royal than the king; it had to be more sober and down to earth than any of the other sciences.

More sceptical than the sceptics

This, in fact, was a problem that beset the SPR. In order to prove
their scientific credentials, many of its members went out of
their way to *disprove* paranormal claims, requiring conditions so
stringent that practically no medium could meet them. Koestler,
who died in 1983, left his estate, valued at close to a million
pounds, to any British university that would establish a Chair
in Parapsychology. The University of Edinburgh did in 1985,
through the help of parapsychologist John Beloff, and the Koestler
Parapsychological Unit is still in operation today.[16] Caroline Watt,
who holds the Koestler Chair in Parapsychology, agrees with Jessica
Utts that there is a 'small' but 'statistically significant' precognition
effect in humans.

The flipside of this was that as parapsychology muffled itself in
Zenner cards, random number generators, and an 'almost fanatical
devotion to statistical methods', creating a 'dreary', 'antiseptic'
atmosphere, the exact opposite was happening in physics.[17] There,
'the deeper the physicist intruded into the realms of the sub-atomic
and super-galactic dimensions, the more intensely he was made
aware of their paradoxical and common-sense defying structure.'[18]

Parapsychology and physics seemed to have exchanged places: the
strange world of precognition and telepathy had become as exciting
as life insurance, while physics, that bedrock of common sense, had
taken a quantum leap into Wonderland, with Heisenberg, Bohr,
Schrödinger and Co. as so many Alices, encountering particles that
acted like waves, and solid matter made of mathematics more than
anything else.

CSICOPS

The two should have met and embraced. Yet the opposite happened.
Irritated by the popularity of ESP, and the general 'occult revival'
of the 1960s and 70s, some hardnosed critics dismissive of the
paranormal, formed a committee determined to undermine what
they saw as parapsychology's pretensions to scientific respectability.[19]

Famously, in 1979, at a meeting of the American Association for the Advancement of Science, the physicist John Wheeler, a leader of the committee – and coiner of the terms 'black hole' and 'wormhole' – raised the battle cry: 'Drive the pseudos out of the workshop of science!'

The irony here is that Wheeler is the author of a theory of reality that in essence is no different than what many mystics and occultists have said for ages: that the world we see is created by the mind. In *The Caretakers of the Cosmos* I point out that Wheeler's 'Participatory Anthropic Principle', which states that human beings are 'participators' in bringing the universe, not only of the present, but also of the past 'into being', was stated a century earlier by Rudolf Steiner when he said that 'Man ... himself cooperates in bringing the world into existence.'[20] I also point out that Owen Barfield, a follower of Steiner and an important thinker in his own right, developed a notion of 'participation' in which human consciousness plays an active role in bringing reality into being, decades before Wheeler. We could push this back even further, and remember that William Blake knew that 'The sun's light, when it unfolds it, depends on the organ that beholds it.' But then Wheeler was a physicist. Blake was a poet, Steiner an occultist and Barfield a 'Coleridge looney' as an academic friend once described him to me. It seems that if you preface your remarks with the word 'quantum', anything is possible, but if you breathe any mention of 'paranormal' you are a 'pseudo'.

Yet it is surely good news that science is coming around to accept precognition – or at least that there is sufficient statistical evidence to support its acceptance. But then it seems that in the past, there already was sufficient evidence for this and for other paranormal phenomena. More than half a century ago, Louisa Rhine wrote that 'a backlog of experimental findings has now accumulated which clearly shows that information can upon occasion get into the mind by ... extrasensory channels.'[21] H. J. Eysenck, a tough-minded psychologist and critic of parapsychology, said much the same. Unless we accept the existence of a 'gigantic conspiracy involving some thirty University departments all over the world,' Eysenck wrote in 1957, 'the only conclusion the unbiased observer can come to must be that there does exist a small number of people who obtain knowledge existing in other people's minds, or in the outer world, by means as yet unknown to science.'[22]

So perhaps it is not the evidence per se, but the attitude of the individual considering the evidence, that counts. With that in mind, let's consider some evidence.

Statistically speaking

In 2011, Daryl Bem, a professor of psychology at Cornell University, published a paper in the prestigious *Journal of Personality and Social Science* that offered impressive evidence of precognition. He had designed several tests to see if behaviour in the present could be accounted for by knowledge gleaned from the future, what J. B. Priestley had christened FIP, 'future influencing the past'. Priestley gives as an example of this the odd feeling of excitement and expectancy he got whenever he received perfectly formal letters from the head of a department he was in correspondence with and whom he had never met. It later turned out that he would marry the department head, and Priestley believes he was getting advance notice of their happiness together. Bem's experiment involved something less charming: multiple choice tests. The idea was that if you could look ahead a day, you could see the answers and then do well. I can't go into the details, but Bem's results showed that indeed this was the case. And what was more, as Radin points out, the results proved to be repeatable.[23]

Repeatability is the bugbear of parapsychology. The fact that most 'live' manifestations of the paranormal are not repeatable – because they are often triggered by crisis and powerful emotion, or, conversely, deep relaxation – is what has excluded them from scientific respectability. But that is like asking someone to fall in love on demand. Such demands have made some explorers of the paranormal, like Stan Gooch, declare that 'The compulsive drive for repeatability' has been 'the ruin of the parapsychologist', and that 'science cannot be of any real use to us in the study of the paranormal.'[24] We may not agree with Gooch, but we get his point. Yet even in the sedate setting of a classroom, the statistics indicated positive results. Radin conducted his own tests for precognition via the internet and received similar results.[25]

Radin also points out the hostility such results prompt in the scientific community. So although the statistics show that the results would be accepted if they were about something else, they still aren't. As the psychiatrist Karl Menninger pointed out, 'Attitudes are more important than facts', a dictum that can be put to dubious use, as I show elsewhere.[26] In this case the attitude is 'Damn the facts! Precognition – and other forms of the paranormal – can't exist.' And so, as Jessica Utts lamented, 'Most scientists reject the possible reality of these abilities without looking at the data!'[27]

Time loops

Another recent work offering scientific evidence for precognition is *Time Loops*, an ambitious and fascinating, if ultimately unconvincing (to my mind at least), attempt to account for precognition from a 'physicalist' basis by the science writer Eric Wargo. After the seminar with Dean Radin, I spent a week in Montreal reading *Time Loops*, and my notebook shows my engagement with it. Among the evidence it offers is a series of tests carried out between 1976 and 1999 at the Princeton Engineering Anomalies Research laboratory (PEARS) involving 'precognitive remote perception', a sort of 'two for one' experiment, combining precognition with remote viewing. Subjects were asked to report their impressions of geographical targets that would later be visited by the experimenters. Precognition came in because 'the target for a given trial was always selected at random after the trial ended.'[28] So, a subject was asked to visualise some destination, and only after he or she had, was a target randomly chosen (this would rule out telepathy as a factor). As in the multiple choice test, the idea was to see if the subjects could get the correct answer in advance. When the results came in from 653 trials involving 72 participants, they were impressive. A third of the trials proved accurate, with odds against chance a considerable 33 million to one.

Another series of tests, some 309 of them conducted between

1935 and 1987, and involving more than 50,000 participants, was subject to 'meta-analysis' using computers, the same procedure as with the PEARS experiments. The results here against chance were 'astronomical: on the order of ten septillion to one.'[29] Along with repeatability, chance is parapsychology's other *bête noire*.

When Pavlov met Minkowski

Yet while this and the other statistical evidence 'proving' precognition provided in *Time Loops* is indeed convincing, I can't say the same for the book's complex argument for how and why it works. The theory is an ingenious blend of the mathematician (and teacher of Einstein) Hermann Minkowski's 'block universe' – a four-dimensional ever-present 'now' that we segment into an illusory past, present, and future – Pavlovian conditioning, Freud, and Jacques Lacan (himself not the most comprehensible of thinkers) – that had me scratching my head more than once. My fundamental objection was that none of my precognitive experiences seem to match up to the theory. Part of the argument is the idea that precognition is a matter of 'producing a behaviour that is tied to a forthcoming reward', which means it is a weird form of operant condition, in which my brain in the future 'rewards' me for my behaviour now. One of these 'rewards', especially in the context of precognition involving catastrophes, is the self-serving reflection that 'I survived'.[30]

This 'better them than me' sentiment seems to disparage human psychology. Forget precognition: when I hear of a disaster on the news, do I really feel 'better them than me'? More to the point, I can't say that any of my precognitive dreams fit this bill, either as rewards or as 'future learning experiences about disaster and death'. The only disaster I dreamed of precognitively was a volcano in Japan, that affected me in no way and which, when I learned about it, I reacted to neutrally. Perhaps because the book seems concerned with establishing 'retro-causation', a kind of cause-and-effect in reverse, as a kind of physical 'law' – making Priestley's FIP as fundamental as gravity – and hence wants to move precognition 'out of the murky realm of the "occult" or "supernatural" and into the realm of physical plausibility', that I feel out of sympathy with it.[31] I see nothing particularly murky about the occult or the supernatural, and have written books explaining why; nor do I see any need for physical plausibility. To relocate precognition in a more scientifically acceptable neighbourhood, Wargo dismisses both 'Victorian superpowers' of the kind that the SPR investigated, and Jung's synchronicity, both of which I am perfectly happy with. This suggests we may not be talking about the same thing.

Yet the statistics are in and can't be swept under the reductive rug.[32] But while some open-minded scientists may be convinced, most of us will raise an eyebrow and simply carry on. What moves us about precognition is the anecdotal material, the sort that most scientists peremptorily ignore. This should keep them busy, because there is a lot of it.

Place your bets

Anyone who bothers to spend a year reading about precognition, as I did preparing for this book, will discover that there is a glut of examples of it. Much of it does involve death and disaster. When I mentioned to friends that I was working on a book about precognitive dreams, I was treated to more than one example from their experience. A standard story was of someone having a dream about a friend who was in danger or ill or in some other precarious situation and then learning soon after that the friend had died or was in hospital or in an accident. That was often the single paranormal experience that person had. When I posted on Twitter that I was working on this book, correspondents sent me accounts of their precognitive dreams; sadly, space doesn't allow me to include them here. Indeed, I can only skim the surface of the huge number of precognitive events I have read about, and the reader must be satisfied with a sampling. As this book is about my precognitive dreams, I shall keep my sampling to other precognitive dreams, although even here there is an embarrassment of riches.

One example concerns the lucky John Godley, later Lord Kilbracken. In 1946, Godley, then at Oxford, woke up from a dream in which he heard the names of two horses, Bindle and Juladdin. Godley checked the racing sheets and saw that both horses were running in different races that day. Godley placed his bets and won more than £100, a considerable sum then. Not long after, he dreamed of a winning horse named Tubermor. There was a Tuberose running that day; Godley took the plunge and won again. Another dream had him calling his bookmaker, who told him the name of another winner, Monumentor. Godley saw that a horse called Mentores was running that day; he backed it and again won. More winners followed. Godley ended his streak of dream winners spectacularly in 1958 when he dreamed that the Grand National was won by a horse named What Man. The real winner was Mr. What, who earned Godley £450, nearly £10,000 in today's money. On the strength of his fame as a 'psychic punter', Godley became the racing correspondent for the *Daily Mirror*, then later for the *Daily Express*, and he wrote a book about his precognitive dreams, aptly entitled *Tell Me The Next One*.

Another dream winner was Wilbur Wright, who wrote crime fiction under the name David Graham. As a RAF pilot during the Second World War, Wright was stunned when his friend and fellow pilot handed him a package of his possessions and asked that he made sure his family would receive it. His friend had dreamed of his death and was convinced it was unavoidable. Wright said he was talking nonsense, but later that day he saw his friend's plane burst into flames and watched as he piloted it into the German airfield they were strafing. After the war, Wright started having his own precognitive dreams, not about his death, but, as John Godley did, about the races. Yet unlike Godley, he was not a punter.

Wright had the same dream three times, in 1946, 1948, and 1954, and each one named a winner. The dream found him at a racecourse, where he asked someone who had won the race. In 1946, his dream friend told him it was Airbourne. Wright said 'But there's no such horse running.' The dream friend said 'Well, it won anyway.' Wright mentioned the dream to some friends who naturally thought he was crazy. He wasn't a betting man himself, so when he saw that Airbourne was running that day in the St Leger, he didn't bother to place a bet. Understandably, when Airbourne won, Wright's friends kicked themselves. In 1948, in the same way another dream informed Wright that Arctic Prince would win the Derby. This time his friends took no chances; they backed Arctic Prince, who was running, and won a tidy sum. Wright again didn't bother to place a bet. In 1954, Wright dreamed yet again of being at a race-course. Yet when his dream tout appeared, Wright realised he was dreaming – that is, it became a lucid dream. He said to his dream friend 'Oh no, not you again!' This was a mistake; he could see that his dream friend was annoyed. Nevertheless, the procedure was the same, and this time the horse was named Radar. No Radar was running, but a horse named Nahar was in the Cambridgeshire. Again Wright told a friend, who placed a bet and won. Wright once again didn't bother. This was the last future dream; as Colin Wilson suggests, Wright's 'dream tipster' was probably put off by his 'Oh no, not you again!' and probably thought it a waste of time to provide winners to someone who couldn't be bothered to bet.[33]

Disasters on canvas

Other precognitive dreams concern less profitable encounters. On September 11, 1996, the London artist David Mandell woke from a nightmare in which he saw two tall towers collapsing with the impact of an earthquake. He had the same dream six months later. This time he painted it, producing a watercolour in which the towers appear with a smaller pyramidal building. Nine months later he had a similar dream, in which two airplanes hit two buildings from opposite directions. In 2001, when he saw the attacks on the World Trade Centre on television, five years to the day of his initial dream, Mandell was shaken. He realised that his watercolour 'exactly matched the New York skyline with the burning towers flanked by the pyramid-topped American Express Building.'[34] This was not the only time Mandell had painted his precognitive dreams, and in 2003 they were the subject of a documentary.[35]

Aberfan

Probably the most well documented precognitive dreams about a disaster concerned the catastrophe at Aberfan, South Wales, in 1966. On October 21 a huge coal slip descended a mountainside, engulfing houses and a school. Of the 144 people who died, 116 were school children. A psychiatrist, John Barker, treating the villagers, had an interest in the paranormal. He convinced the *Evening Standard* to ask those of its readers who felt they had had a premonition of the disaster to write in about it. It turned out that at least 76 people had premonitions, and of these 36 came in dreams. As Steve Taylor points out in *Making Time*, some dreams were so frightening they woke the dreamers.[36] Eryl May Jones, a ten-year old student at the school, told her mother the day before the landslide that she had dreamed of 'something black coming down' and covering the school. She was one of the victims. The number of dreams and their accuracy led Barker to establish the British Premonition Bureau that year, with the hope that future catastrophes could be avoided if dreams and other presentiments of danger could be recorded in advance. The project lasted five years and was not considered a success.[37]

During the Covid-19 pandemic, after tweeting about my dream from 1998 telling me to 'Just stay home', I suggested that 'It would be interesting to collect dreams from before coronomania, to see if there were any indications', citing the British Premonitions Bureau.[38]

Random precognitions

Other precognitive dreams have an air of inconsequence that suggests that whatever is responsible for them acts randomly or has an odd sense of humour. Here are a few from the many accounts in Brian Inglis' *The Power of Dreams*.

In *Second Sight in Daily Life*, Waldo Sabine recounts a dream in which an Indian fakir shows him a map. That evening he saw the same map at a lecture on psychical phenomena. A Mrs Atlay dreamed about seeing a pig in her dining room while she was reading prayers. Her family laughed when she mentioned it. Yet, lo and behold, when she opened the dining room doors after prayers, she indeed found the pig waiting for her; it had escaped from its pen and got in by a back door. In *Imagination in Dreams* Frederick Greenwood recounts a dream in which he found himself touching 'a woman's hand neatly cut from the wrist.' Later that day, on a business call, he found himself touching the hand of a female mummy, resting on a mantelpiece, a macabre bit of bric-à-brac. A BBC engineer dreamed of a sparrow hawk resting on his shoulder. That morning, for laughs, one of his housemates sneaked up on him and perched a stuffed sparrow hawk, left out for the rubbish, on his shoulder; he hadn't mentioned the dream. A woman dreamed of her younger sister reaching into her coat pocket and pulling out a handful of bottle caps. Later, at dinner, wanting her cigarettes, her sister reached into her coat pocket and discovered the bottle caps, put there by her young sons.

I could go on. These and other examples suggest, as Inglis remarks, that 'Sometimes it is as if the practical joker lays on the dream to confound the sceptic.' Some are not amused. Charles Dickens dreamed of a woman in a red shawl who introduced herself as 'Miss Napier'. Later that day at a social event, he met a group of women. One of them, wearing a red shawl, was introduced to him. Her name? Miss Napier. Dickens' response to this was not astonishment, however, but annoyance. What did this Miss Napier, who would play no role in his life, have to do with him? The philosopher Schopenhauer's reaction to a similar experience was different. One morning Schopenhauer spilled some ink and called his maid to clean it up. She remarked that she had dreamed this the night before. Schopenhauer scoffed, but when her fellow servant confirmed it as true, he reflected that 'everything that happens, happens of necessity', a fatalism one can detect in his philosophy.

A dream in time

Some precognitive dreams save lives. On holiday in Atlantic City, the suffragette Susan B. Anthony dreamed that she had died in a fire. She took the hint and that morning returned to Philadelphia. Later that day her hotel and those nearby burned down. Some dreams, however, prove no defence. Three days before his assassination, Abraham Lincoln dreamed of wandering through the White House, seeking out the source of the woeful sobbing he heard. When he entered the East Room, he saw a corpse guarded by soldiers and a crowd of mourners. This was not unusual, as Lincoln had an interest in premonitory dreams.

On June 28, 1914, Archduke Franz Ferdinand was assassinated in Sarajevo, Bosnia, igniting the First World War. At 3:45 that morning, Bishop Joseph Lanyi, the archduke's old tutor, woke from a dream in which he was looking at a letter, written in the archduke's handwriting, and bearing his black seal and coat of arms. Included was a picture that showed the archduke and his wife, the Duchess of Hohenberg, in a car on a street near a narrow passageway. There were crowds and the bishop saw two men emerge from them and fire at the archduke. The text of the letter informed the bishop that the archduke and his wife would be assassinated that day. It ended with the date and time, 3:45, and was signed by the archduke. The bishop recorded the dream in detail, and even drew the scene he had witnessed. His account and drawing were certified, and sent to his brother, a Jesuit priest. The one discrepancy was that there were two assailants in the dream, while Gavrilo Princip was the sole assassin.

Dunne and his dreams

Again, this is only a small sampling of the many accounts of what we can call 'Dunne-type dreams'. Given that, what did J. W. Dunne himself think about his future dreams? In *An Experiment With Time* Dunne remarks that his dream about his watch stopping at 4:30, which started his 'experiment', was a 'peculiar one (in ways which have nothing to do with this book).'[39] This seems an odd comment; isn't the dream's peculiarity that it was precognitive? What did Dunne mean? It wasn't until after Dunne died that his readers learned exactly how peculiar his dream was.

In *Intrusions*, published by his wife in 1955, six years after Dunne's death, Dunne reveals the peculiar circumstances around his first 'Dunne-type dream'. He recounts how in his youth he was interested in spiritualism and how at a séance the 'spirit guide' of the medium had announced that 'there is a young man here tonight who will be the greatest medium *that the world has ever seen.*'[40] That young man, Dunne was surprised to discover, was himself.

Yet he could not have been that surprised; since he was a boy, Dunne had had the conviction that it was his destiny to bring an important message to humanity. His posthumous work, *Intrusions*, was an account of how, at different times, unknown agencies had 'intruded' in his life in order to help him fulfil that destiny. They had even intervened to ensure that he would write his last book. Dunne told a correspondent of how, during the Second World War, he put off writing it, in order to work on a new aircraft design. But a case of pneumonia led to pleurisy in his ribs, and added to this was an attack of neuritis in his right arm. The two made it impossible for him to work at his drawing board, so he returned to *Intrusions*, which, sadly, he didn't live to finish.

In a dream at this time he heard voices saying 'Hurry! Hurry!' Given Dunne died before he could complete his last work, the urgency was apt. In the same dream, an angel told Dunne that it would be a 'two-thousand-years long calamity for mankind' if he didn't finish the book, which was intended to clear up some obscurities in his ideas. What Dunne had not mentioned in *An Experiment in Time* was that he had heard these voices earlier, when he had his first future dream. Then they had said 'Look! Look!' and it was because of this prompting that he got out of bed, lit a match, and looked to see that the watch had stopped at 4:30, as it had in the dream. Something or someone was making sure that he would pay attention to this dream – to, in fact, see that in it he had dreamed the future.

The reason Dunne didn't want to mention the voices in his dream or his early interest in spiritualism, is that he didn't want the message he had to convey to be tainted with any stain of 'occultism'. That would have made it easy to ignore. A reader of Dunne soon sees that for him his dreams were merely the medium through which he came to understand the strange character of time. He does not seem to be interested in dreams for their own sake, and there's no evidence he was aware of the literature on dreams, although he does mention psychoanalysis here and there. What Dunne focuses on in the books that followed *An Experiment With Time* is what he called 'Serialism', his theory of 'serial time'. Dunne believed that his theory of serial time 'proved' human immortality and the existence of God, or what Dunne called 'Universal Mind'. If this is so, we can see that he indeed had an important message for humanity.

Unfortunately, Serialism did not catch on, neither as proof of God or of immortality. Most writers intrigued by Dunne's future dreams, appreciate the dreams but reject the theory; I do myself. And for the average reader, while the accounts of Dunne's dreams are gripping, his theory is less so and requires something of a mathematical turn of mind to follow. T. C. Lethbridge summed up the general feeling when he remarked that 'it would have been better if the theory had not been included' in Dunne's books at all, as 'it did not seem to make the question' of precognitive dreams 'any easier to understand' and even tended to 'obscure the basic fact'.[41]

Serialism

What is Serialism? We've come across a form of it already, when I looked at the idea of time as a 'stream' or 'river', as something that flows. Were I to find a spot on the banks of time, to observe its flow, I would still be in *another* time, from which I could gauge the first time's speed. But this other time itself would flow, and so would require yet another time by which to gauge its speed, and so on. A related question is, how *long* does it take to travel in time? If not instantaneous, then time travel itself takes time, and we find ourselves in the same muddle. This lands us in the philosophical horror of 'infinite regress', rather like the effect of holding one mirror up to another. The reflections, though appearing smaller, do not end.

Dunne tried to explain Serialism in *Nothing Dies*.[42] Say you are interested in forming a picture of the world that will be as accurate and comprehensive as possible. Dunne uses the example of someone painting a picture of a landscape. We see the hills, trees, and a house, yes, but something is missing. What? If we want to have an accurate and comprehensive picture, we have to include the artist himself. This requires another picture, one of the landscape *including* the artist. But who paints this? Shouldn't that artist be in the picture too? All right. But now we need a third artist, painting a picture that includes the landscape and the first two artists. What is happening here? There is always someone *outside* the picture responsible for making it.[43] If we want to catch him, we require yet another artist, and so on.

William James once had an exchange with an elderly lady who disagreed with his account of the solar system. When he politely asked her for her version, she explained that we live on a crust of earth which is resting on the back of a giant turtle. James asked her what was supporting the turtle; she replied that it stood on an even larger one. When James asked what the second turtle rested on, the lady replied 'It's no use Mr James. It's turtles all the way down.' Dunne's serialism doesn't work with turtles, but with different kinds of time and the different 'Observers' in them. With Dunne it's Observers all the way down.

Dunne explained his future dreams by a series of 'Observers' who exist in different times. Observer 1 is the artist painting the picture. Observer 2 is the artist painting the picture that includes Observer 1. Observer 3 paints the picture that includes Observers 1 and 2, and so on. This would continue ad infinitum, except that Dunne argues that eventually, if we trace back our individual self-consciousness, we would arrive at what he calls a 'superlative general observer, the fount of all self-consciousness' a 'tree of which we are the branches', and in which we 'live and have our being.' In other words, the ultimate turtle.

Aristotle proposed a 'first cause' or 'unmoved mover' as the ultimate source of all subsequent 'causes' and 'movements'; Bishop Berkeley, who believed that 'the essence of things is to be perceived', said that the reason things do not disappear when we are not looking at them is that God is always looking at them, and so maintains their existence. We can see Dunne's 'superlative general observer' as a combination of the two. This is the Universal Mind of which we all are parts, the ultimate Observer who calls a halt to the infinite regress.

Observing the observers

How does this account for precognitive dreams? Observer 1 lives in ordinary time, moving from the past to the future. Their job is to gather information, to gain knowledge, and learn about the world. To this end, their awareness is limited to a narrow band of reality, the equable flow of Newtonian time. Here Dunne agrees with Bergson about the brain being the organ of 'attention to life', to dealing with the world (we would say the left brain). When we sleep, our awareness slips out of this harness and enters Time 2, where the past and future are equally available. Hence our dreams are often a muddle of the two. This is why we most often don't recognise the bits of the future in our dreams unless we make an effort to remember the dreams, to record them.

Our immortality comes into the picture because it is only Observer 1, whom I mistakenly believe to be the 'real me', that dies. Observer 2 in Time 2 carries on. In a way we can think of ourselves in Time 1, ordinary time, as on a reconnaissance mission, gathering data to be transmitted to the Universal Mind. It is a grand idea, but the path to it is not as clear or convincing as Dunne would have liked. The idea of an almost infinite series of 'me' halting at an arbitrary 'superlative observer' seems unsatisfying, although scientists prone to accept the Big Bang have to grant at least one 'causeless cause' in order to get the universe going.

Priestley and Ouspensky

Someone who appreciated Dunne's work but didn't care for Serialism was the 'time haunted man', J. B. Priestley. Priestley is not read today as much as he should be, but in the mid-twentieth century he was one of the most widely read of English authors. One of his most popular plays, which is still performed today, *Time and the Conways*, is based on Dunne's ideas. Priestley was also deeply influenced by the time theories of P. D. Ouspensky. Oddly, like Dunne, Ouspensky was interested in both time and dreams, but he didn't make the connection between the two, although Ouspensky did experience precognition in his youth and also during a series of experiments in altering his consciousness, which he wrote about in an essay called 'Experimental Mysticism'.[44] Ouspensky's notion of 'eternal recurrence', the idea that our lives are lived over and over innumerable times, informed Priestley's play *I Have Been Here Before*. Ouspensky accounted for his frequent experiences of *déjà vu* and his early precognitive experiences in terms of recurrence. It's not known whether Ouspensky knew of Dunne's work; if he had he might have recognised that precognitive dreams offered another explanation.

Although he made use of Ouspensky's 'eternal recurrence', Priestley did not accept it, and was more at home with Ouspensky's notion of a 'three dimensional time' to complement our three-dimensional space.[45] Ouspensky's three temporal dimensions line up nicely, if not exactly, with the three 'times' or 'Observers' to which Priestley limited Dunne's infinite regress.

As Anthony Peake shows in his study of Priestley, his odd relationship to time started at an early age. He may even have been saved from an early death by a premonition.[46] When the First World War broke out, rather than join his local West Yorkshire regiment and become one of the 'Bradford Pals' – Bradford was his home town – Priestley joined the West Riding regiment. He called his odd decision a 'signal from the unknown'. Whatever prompted him, it proved a good choice. Some 1,777 of the Bradford Pals were killed within the first hour of the Battle of the Somme. Had Priestley been among them, he most likely would have joined them.

Priestley had other precognitive experiences. When he was a boy, he dreamed that his favourite uncle 'appeared in the doorway, very angry', glaring at him. Years later, on leave during the war, Priestley suddenly saw his uncle, 'across the length of a bar, and there he was, very angry, glaring at me.'[47] On holiday in Italy, Priestley sat by his wife, who was having a nap. He drifted into a doze – the hypnagogic state – and he then 'saw her open her eyes, smile slowly … yawn and stretch.' A few moments later, when he was 'wide awake', she did exactly the same thing.[48] In a peculiar dream he saw people wearing masks with 'moveable mouths'. Years later these masks turned up as props in his play *Johnson Over Jordan*.[49] His most spectacular future dream involved the Grand Canyon. Sometime in the 1920s, Priestley dreamed of being 'in the front row of a balcony or gallery of some colossal vague theatre.' On the vast stage he saw 'a brilliant coloured and fantastic spectacle, quite motionless, quite unlike anything I had ever seen before.' Years later, during his first visit to the Grand Canyon, Priestley found himself sitting before the railing in front of his hotel, waiting for the early morning mist to clear. When it did he saw 'as if I were sitting in the front row of a balcony, that brilliantly coloured and fantastic spectacle … I had seen in my dream theatre.'[50] Priestley had dreamed of seeing the Grand Canyon a good decade before he went there.

Years later, when writing *Man and Time*, Priestley appeared on the BBC television programme *Monitor* to talk about the book. The interviewer asked the audience to send in any accounts of precognition or other unusual experiences of time. The response was overwhelming. Priestley was flooded with letters, just as Brian Inglis was when, in an article he wrote about dreams in 1985, he asked for readers to write to the Koestler Foundation about any odd dreams of theirs. Clearly there is a large public interested in these things, whatever the experts say.

Priestley presented some of the many accounts of precognitive dreams that he received in *Man and Time* and also in a later book, *Over The Long High Wall*. This book was the 'supreme flop' of his career, Priestley says; if he is still listening in to events on this plane, he might be cheered to know that I have read it several times and find it fascinating. Space does not allow me to present many of the accounts Priestley received. They would not, in any case, add more to our anecdotal material. But one dream in particular helps us understand how he understood Dunne's ideas. Priestley tells us about a woman who dreamed that she was at a creek with her baby, and wanted to wash some things, but had forgotten the soap. In the dream she returned to her tent to fetch the soap, and when she returned to the creek, her baby was dead in the water. When she later found herself at a creek on a camping trip, and, as in the dream, wanted to wash some things, she remembered the dream, and decided to bring the baby with her.

Was the dream a warning? As Priestley says, 'The future can be seen. And because it can be seen, it can be changed.'[51] How was it changed? Through Time 3, or the third dimension of time, which is one of freedom and creative action.

As easy as 1, 2, 3

Priestley saw that to account for his own precognitive dreams and other odd experiences of time, he did not need Dunne's infinite regress. All that was necessary were the first three Observers in the first three times. If we look at our own experience we can see what he means. Observer 1 is me moving along in everyday, tick-tock time, my attention fixed on the present, whatever it is that I have to deal with at the moment. Observer 2 is that 'me' who opens up in moments of relaxation, or in hypnagogic states, who is aware, as Colin Wilson says, of 'other times and places', who does not have his attention rigidly fixed on the present moment, or conversely, whose present moment 'expands' a bit to include more of reality. This is the 'me' that in dreams has access to the past and the future.

Observer 3 is, in a way, *not* an observer, at least not solely one, but is capable of action. As Priestley writes, 'whenever I am offered this block universe' – Minkowski's ever-present 'now' – 'all solidly there to be discovered, in the light of our successive moments, I am never called anything but an Observer.'[52] He is, he writes, 'like a man on a moving roadway in some world's fair.' For Priestley, this 'block universe' is not so adamantine as the mathematicians and physicists who propose it believe. For Priestley there are two 'futures'. There is a 'future already existing so that it can be discovered by one part of the mind' – the dead baby – and a 'future that can be shaped by the exercise of our free will' – the living baby the mother saves from drowning.[53]

Observer 3, then, is the 'me' who can *change the future*, who can not only step back from the press of time and events and, in dreams and other moments, get a glimpse of the future, but who can act. Which means that this 'me' possesses free will, something that precognition seems to deny. It is also in this Time 3 that Priestley places the acceleration of time he experiences in some creative bursts. He speaks of writing some of his plays at a breakneck speed he would be unable to maintain in Time 1.

Your life is your time

Ouspensky's three dimensions of time are not the same as Dunne's Observers, but in terms of our experience they are very close. The first dimension of time is our usual, everyday, Newtonian time. The second dimension is that of recurrence, which we can see as the 'eternity of each moment'.

As we move along our timeline, we leave moments behind and tend to believe that they simply fade away. Ouspensky doesn't agree. Each moment of passing time exists in its own 'eternal now'. For Ouspensky, time isn't a river or stream in which we exist for a while, then disappear. A person's life *is* their time; there is no time for them outside of their life. There is, then, no abstract Newtonian time; or at least we each of us have our own time, running alongside it. We are not in time; time is in us. If time is a river, then each of us carries our own inner tributary. As Ouspensky writes, '*Life* in itself is *time* for man ... there is not and cannot be any other time outside the time of his life. *Man is his life*. His life is his time.'[54]

This was an idea Ouspensky gave fictional form in his novel *Strange Life of Ivan Osokin*, in which the hero, faced with a life of failures and mistakes, asks a magician to send him back in time, so that he can change things.[55] Ouspensky believed in a strange form of reincarnation, in which we do not reincarnate in the future, but in the past, so that we can attempt to alter what has happened, to make good what had gone bad the first time around. This we know is the theme of dozens of science fiction films, such as *The Terminator* and *Twelve Monkeys*.

There are logical problems in this idea of reincarnating into the past, as there are with the notion of our life being our time. If, when I die I immediately return to the moment of my birth, as Ouspensky says we do, how do I account for people in my life who continue living, yet who must 'recur' with me, as they are players in my life story? And isn't the attempt to return to the past in order to change it – to, as Ouspensky says, defeat present evil by uprooting it in the past – subject to the same paradoxes we found in the notion of literal time travel? But we shouldn't let these conundrums obscure the essential significance of Ouspensky's ideas. Because his third dimension of time is one in which all the other possibilities of each moment are available to be actualised. In this dimension of time, we can take a different path when we reach our crossroads.

At each moment, we face a number of possible choices. Some are trivial: what to eat, what to wear, what to watch on television. Others are more significant: whom to marry, what work to do, whether to devote our life to personal ends or to something more profound. In tick-tock time we make one choice; in recurrence that choice becomes eternal. But in the third dimension of time, the other possible choices are still there, and, as we saw with the woman who did *not* leave her child by the creek, although she dreamed that she had, we have the power to actualise a different possibility. This is an idea that has had a quantum brush up and is now presented as the 'many-worlds theory', with parallel universes – possibly an infinite number of them – attached to ours, vibrating at a higher rate, and so invisible to us. This is again an idea associated with John Wheeler, and again an 'occultist' was there before him. The entire occult or esoteric tradition has always spoken of realities other than our everyday one. About time the physicists cottoned on to it.

It takes some effort

So, for Ouspensky and Priestley, there is a kind or dimension of time in which we can act to alter things, in which we are not stuck with the fate that seems laid out for us, but can take steps to change it. But it does not happen automatically. Automatic or mechanical living just happens; we plod through life, taking what comes, making no real effort to exert free will, which is not the same as having things 'your way' or following your 'true will'. Because that is the essence of it: effort. Effort, to me, is proof positive of free will. No machine can make an effort, nor can any computer, no matter its operating system. Only a being with free will can decide to 'pull themselves together' and 'try harder', can strain to reach reserves of energy and power *they very often don't even know they have*. What machine can do that?

Although Ouspensky believed that many, if not most people's lives simply recur, there was a way to prevent one's tomorrow from simply being a repeat of today. There was a way to *remember* and so to be aware of other possibilities. For many of us, recurrence happens every day; we go through the same routines, think the same thoughts, feel the same feelings as we did the day before. It is usually only in times of crisis or some unusual circumstance that we actually *feel* ourselves as free, in the real sense, not a political or social one, although of course those are important. What I mean is that we actually feel *alive*. When we feel this we are in the third dimension of time, we are Observer 3, awake in what Maurice Nicoll calls 'living time', and Priestley in his novel *The Magicians* calls 'time alive'. Ouspensky taught a particular method of achieving this feeling and passed that on to Nicoll. There are others. In a sense the method isn't as important – all roads will take you somewhere – as the knowledge that this freedom, this 'aliveness' is real. It is, in fact, the realest thing about us.[56] It is for this reason that Priestley insists that 'what is important is how we live here and now, what we can make while we are still able to be makers.'[57]

T. C. Lethbridge

Someone else who was fascinated by Dunne-type dreams but not Serialism was the Cambridge don T. C. Lethbridge. Like Priestley, Lethbridge should be read more than he is these days. Lethbridge, an archaeologist, was for many years the Keeper of Anglo-Saxon Antiquities at the Cambridge Museum of Archaeology and Ethnology. After retiring from Cambridge in 1956, Lethbridge began a second career as a parapsychologist and explorer of mysteries, writing a series of very readable and insightful books on witchcraft, UFOs, ESP, ghosts, ancient religion, megaliths, and other strange phenomena.[58] He is most known for his work with pendulums, which grew out of his study of dowsing. Dowsing is the ability to detect hidden objects through the use of a dowsing rod, most often a forked twig, which is held in the dowser's hands; it reacts by shaking involuntarily when it has located the object. It is usually associated with locating water and is used by most professionals in the field, although they are often loathe to admit it.

Lethbridge discovered that a pendulum works just as well and can be used to locate not only water, but practically anything. It can even work over a map. Through a series of experiments, Lethbridge discovered that the pendulum would respond not only to physical objects or substances, but also to abstract ideas and thoughts.

One of Lethbridge's discoveries using the pendulum was that different substances had their own 'rate'. This was the length of the string holding the bob when it would react to the substance. So, for example, Lethbridge discovered that the rate for silver was 22 inches (56 centimetres). Lethbridge had devised a simple method of changing the length of the string. He took a small hazel wood ball, drilled a hole in it, inserted the string, knotted one end, and then wound the rest of the string around a pencil. He then stood over whatever substance he was experimenting with, and slowly unwound the string. The pendulum would swing back and forth until it 'found' the substance. Then it would suddenly swing in a circle. For silver, this happened at 22 inches. So 22 inches was the 'rate' for silver.

Lethbridge also discovered that silver shared this rate with other substances, lead, calcium, sodium, but also with the colour grey, which isn't a substance (exactly what colour is remains a mystery, even with Newton's optics). Lethbridge found that he could tell the difference among these things by the number of times the pendulum swung in a circle. For each substance it was different; for silver it was 22:22, for the colour grey it was 22:7. Lethbridge discovered that there was a correspondence between the things that shared the same rate. Forty is the rate for 'black, cold, anger, deceit, sleep and death'; life, which has a rate of twenty, shares it with 'the colour white, and also with the earth and electricity.'[59] These different things seemed to 'go' together. It seemed that Lethbridge was discovering support for the ancient idea of the 'sympathy of all things', the notion that things are related in ways other than cause and effect, through *meaning*, something we will return to.

The next whorl of the spiral

I can't go into all the details of Lethbridge's discoveries – the reader is advised to read him at the source. But one discovery that relates to our subject is that 40 inches (101 centimetres) seemed to be the limit of his rates, at least in our physical world. Nothing he tested reacted to any lengths beyond that. Even abstract ideas didn't, especially one, time. Lethbridge mapped out the rates of different substances and ideas along a spiral, and he found that in the 'next whorl of the spiral' beyond that of our physical world, in what seemed to be another level of reality, time, as we understand it, didn't exist. As in Dunne's and Priestley's Time 2, along the 'second whorl of the spiral', there appeared to be a 'timeless zone', or at least a zone that was not segmented into past, present and future.

This discovery was helpful when Lethbridge's publisher asked if he would be interested in writing a book about dreams. At first he wasn't interested and claimed that he rarely remembered his dreams. But he did remember a curious incident when he was interviewed by the BBC about his work with dowsing. When the cameraman arrived at Lethbridge's home, he looked stunned. Lethbridge watched as he looked around with an odd dazed expression. Finally Lethbridge asked the cameraman if he had been here before. He replied that he had, not in real life, but in a dream, actually in a series of them. It turned out that he had on several occasions dreamed of Lethbridge's house, but not as it was in 1964, at the time of the interview, but as it had been in 1896. He remarked on changes that had been made to the grounds and buildings since that time, and a look at the plans showed that he was correct. So, the cameraman had had a precognitive dream about visiting Lethbridge's house at a time before Lethbridge lived there and the cameraman was even born.

A classification of dreams

This got Lethbridge thinking. He had read Dunne's book and decided to follow his lead and record his dreams as soon as he awoke. As I would some years later, Lethbridge discovered that he, too, dreamed the future. He woke from a dream in which he saw the face of a man he didn't know, framed by what seemed a mirror, and making movements with his hands in front of his face.

Lethbridge thought he was shaving. Later that day as he was out driving, Lethbridge turned a bend and saw the man in the other lane. The windshield framed his face as it was in the dream, and the movements Lethbridge took to be him shaving were actually him at the steering wheel. In the appendix to his last, posthumous book, *The Power of the Pendulum*, Lethbridge provides more examples. As he was waking one morning he had a hypnopompic vision of what he took to be a postcard of the 'black painted hull of a coal-burning steamship ... a passenger vessel with white upperworks.' Lethbridge, an experienced seaman, thought the ship had been built some time ago. Soon after the dream, he heard on the news that the old Queen Elizabeth was being taken out of service. In another vision he saw a large wooden wheel with many spokes and the bottom of the vehicle to which it was attached. That afternoon he saw the same wheel in a photograph of an old 'boneshaker' bicycle.[60]

Lethbridge classified his dreams, in a similar way to the classification Frederik Van Eeden had produced earlier in the century. There were 'flash dreams', pictures that suddenly appear with no narrative or story. 'Strip dreams' were not about striptease; Lethbridge actually believed we had much fewer erotic dreams than is usually assumed. They were a kind of flash dream in motion, like watching a brief clip from a film. Telepathic dreams came from someone else's mind, something J. B. Priestley and William James had experienced. 'Shock dreams' were what caused sudden body jerks as we fall asleep; I often have a hypnogogic vision of a ball being thrown at me as I nod off, and I jerk myself out of its way. I've already noted his distinction between the dream that is really a kind of 'drowsy thinking' and the 'true dream', which comes from some deeper source.

One of Lethbridge's oddest types of dreams is the 'backward dream'. In a dream he saw 'something like a brown, furry snake coming into the room beside my bed.' This snake was followed by 'a brown lump attached to it.' Then he saw that 'a complete brownish cat had come into the room backwards.'[61] Lethbridge reflected that cats sometimes do walk backwards, but it is not their preferred direction. What Lethbridge realised is that what he actually saw was his cat, *leaving* the room, but for some reason he saw it backwards, like a film clip in reverse.

Time and free will

This was one of the oddities of the dream realm, in which we enter the second whorl of the spiral, the timeless zone. Lethbridge was already aware that time is different for different minds. 'Time to a blue tit,' he wrote, 'is quite different from time to an oak or to us' – something with which the Hopi would agree.[62] There is no 'fixed universal scale', no abstract time. What there is, is sequence, in other words, process. In the timeless zone, the future is as present as the past. There, one can 'read the headlines of a newspaper, which will not be printed for days ahead', as Dunne did. We are not in 'earth life' when we do this; we are 'beyond the point of sleep and of death.'[63]

This is reassuring, but it also raises the question of free will. Are 'events on earth no more than a projection' from the next level? Can we change them, as Priestley and Ouspensky believed? Lethbridge leaves the question open. As with Dunne, Priestley and Ouspensky, there is yet another whorl to his spiral, and in the third whorl it does seem that some kind of freedom or creative action is possible. Lethbridge believed that our mission here was to learn as much as possible, to enlarge our minds, in a word, to evolve, to expand our consciousness. Like Dunne he believed in a higher mind, the 'superconscious mind', which is as above our ordinary waking consciousness as our subconsciousness is below it. The superconscious mind is what sees the future, yet Lethbridge argues that it may well be our own mind, on another level of vibration, another whorl of the spiral. He even suggests that it is 'what we are going to become when we have finished our lives on earth' and what we already are 'in deep sleep', the place of true dreams.

Chapter Five
What a Coincidence

Synchronicities and precognition. Helmut Schmidt. Magic and coincidence. Meaning. What is a coincidence? Two things needed. Do they happen to animals? Do accidents happen in nature? Kammerer and coincidence. Degrees of coincidence. The Library Angel. Plum pudding. Melchizedek. Fragrant Floriline. Koestler's ink-fish. Lamarck. Seriality. Law of large numbers. Jung, the I Ching *and synchronicity. 'Herr Doctor Professor.' The coin method. Spiritual agencies. Archetypes activated. Transcendent function.* Unus Mundus. *The Pauli Effect. The brain.*

If precognition presents us with a particular bundle of problems, coincidence, and its peculiarly meaningful form known as synchronicity, has difficulties all its own. It is interesting to compare the two; though different, they are often associated, for understandable reasons. While on the whole precognition is considered frankly impossible, we've seen that it can be tested for by parapsychologists and that the results of these tests show strong evidence for its reality, statistically at least. Coincidences, on the other hand, are accepted as real. No one doubts their existence, but we cannot test for coincidence; at least I can't think of a way in which we could. Coincidences 'happen'; as far as we can tell, we don't make them happen. Coincidences are, as my *Oxford Dictionary* tells me, a 'remarkable concurrence of events' brought about 'apparently by chance'.

Precognition, on the other hand, does appear to be something we do, even if we aren't aware of doing it, and parapsychologists can devise experiments to point this out. So, in the 1970s, the German physicist Helmut Schmidt devised experiments to test if volunteers could predict random sub-atomic events. Over a series of some 60,000 trials he arrived at positive results that were a billion to one against chance.[1] 'Against chance' means that the results were better than could be expected if arrived at randomly, that is, by coincidence. So the tests were actually a means of sequestering coincidence, limiting its importance as a causal factor.

This is true of practically all parapsychological experiments: they aim to show whether some agency *other* than chance was at work in producing the results. If a scientist intent on dismissing precognition wanted to, he or she could say that what Schmidt's results showed was a strange ability in his volunteers to 'create' coincidences, to 'make them happen'. In this case, it would be a coincidence between the guesses of his volunteers and the actual random sub-atomic events. So it could be said that what Schmidt and other parapsychologists had statistical evidence for wasn't precognition, or other paranormal powers, but a peculiar and hitherto unknown ability in some humans to produce coincidences – which, on the face of it, seems rather odd.[2] I would say that an ability to produce coincidences does not seem that distant from what we call 'magic'. And although scientists like Dean Radin want to convince us that 'real magic' exists, most scientists, I think, would find evidence of a human ability to produce coincidences as bad as, if not worse than, evidence for precognition. And when we introduce the element of *meaning*, which turns a coincidence into a synchronicity, and which is as unpredictable an experimental factor as you could get, the difficulties of devising tests for this seem insurmountable.

Just a coincidence?

But what exactly is a coincidence? It is not simply that things *coincide*, that is, that they happen at the same time. Right now I am at my desk, working on this chapter. As I do many other things are happening. I can hear birds outside my window and a bus rolling by; if I turn my head I can see people walking past as well as my neighbour's cat. And of course there are countless other things happening of which I am not immediately aware, from the activity of insects in my garden to that of the nuclear reactions on the sun. But none of these concurrent happenings constitute a coincidence, because none of them are immediately related to me in any way. But if I happen to think of a friend I haven't heard from in a while, and then see that he has just sent me an email or text, or if I am writing about a particular person and hear his name mentioned on the radio at the same time as I am typing it, that would be a coincidence. 'What an odd coincidence,' I would say to my friend when I replied. 'I was just thinking of you when your email arrived.' Or 'How strange,' I might say to myself. 'I was just about to type Bergson when the radio announcer said it.'

Of course if you, like myself, belong to what Arthur Koestler called the secret guild of 'the collectors of coincidences', you would no doubt have noted down several examples of precisely these sorts of coincidences, which are fairly common and happen to practically everyone.[3] My own notebooks are filled with them. For example, while working on this book, I was writing to someone about precognitive dreams. As I did I got an email from my editor, asking for an update on the book. As I copied a link to send to him about my tweet about coming across a dream from 1998 in which I am told to 'Just stay home. There's no reason to go. Just stay home, where it is safe' – precisely the advice we were receiving about how to respond to the coronavirus pandemic – on Radio 3 the BBC announcer remarked about a 'dream team' they had assembled for some project. Earlier, when I began looking through my dream journals and read my comment about the length and complexity of the dreams I had recorded – 'one of those long epics that are really quite tiring' – the Radio 3 announcer said 'pleasant dreams...' These are just two examples of many when what I was thinking or about to write or say was said by someone on the radio.

Many scientists would dismiss the idea that coincidences of this sort have any particular importance; so much goes on around us that it stands to reason that every now and then things sharing a similar character or that seem connected in some way would 'coincide', would happen simultaneously, and appear to be 'meaningful' in some way. Aristotle had said as much. But, these scientists would explain, the meaning is only apparent. It is funny, odd, and worth remarking about – 'Hmm, fascinating' – but in the end the apparent meaning was only a product of chance.

Or was it?

It takes two to coincide

Two things seem indispensable for a coincidence to take place. One is the apparent connection between the things that coincide: my thought of my friend and his email arriving as I think it. Let us put aside for the moment the possibility that I somehow unconsciously knew that my friend had emailed me – leaving aside how I could have obtained this knowledge – and that this floated up to my awareness just as the email arrived. Possibly a telepathic link sent the information on ahead. If my friend had some important news for me, or if he was in some danger or peril, that would suggest something of the sort was at work. But in this case, let's say he just wanted to say hello. There's the connection.

The other thing necessary for a coincidence is you. Coincidences happen to people. Without you or me or anyone else, would there be any coincidences? Do they happen to animals? I don't know. A related question I often consider is: do accidents happen in nature? For something to have happened accidentally means that it wasn't intended. It follows then that for accidents to happen in nature there would have to be intention in nature, that is, purpose or design. If lightning strikes a tree and it crashes on to your house, that is an accident. But if the tree fell in the forest with no human structures around, is that still an accident, or just 'nature'? If an asteroid crashes into the earth, from our perspective that would be an accident on a gigantic scale. But from the universe's point of view, assuming it had one, it would merely be one expression among others of physical laws at work. Most scientists reject the idea that nature has intentions. We read intention into natural phenomena, as when the weather forecaster tells us that the temperature will 'struggle' to rise above single digits. This is an example of the 'pathetic fallacy', attributing human feelings and purposes to inanimate objects. The temperature will not struggle. It is not a single entity and it doesn't 'do' anything nor does it 'want' to reach a certain level. Intention seems a peculiarly human thing and so, I would say, are accidents.

Meaning

Earlier I used 'meaning' in speaking of coincidences. I should have used 'connection' or 'relation' because there is a particular kind of coincidence in which meaning is the central factor. 'Meaning' is another of those terms we use which, as the philosopher Whitehead said, are 'incapable of analysis in terms of factors more far-reaching than themselves.' If something is meaningful we say it is important. Why is it important? Because it is significant. What is significant about it? Well, it *means* something... And so we find ourselves going round and round, never reaching a spot where we can crowbar any of these imponderables up to see what is beneath them. The difference I want to establish between 'meaningful' and 'related' or 'connected' is that in a coincidence, the things coinciding have some kind of correspondence; they are related to each other. So if I am sitting in a café and reading a book and I come across an unusual name and then look up and see the same name on a bus advertisement, the name in the book and on the bus are related, there is a connection between them.

But this fact, though odd, does not necessarily have any particular meaning for me. It is odd, but I am not necessarily moved by the coincidence. Yet, say the name is that of someone I knew and felt I had unintentionally slighted in some way. When I first saw it in the book, it reminded me of this and of the fact that I had intended to redress this mistake but never got around to it. I resolve once again to do this, and as I do, I then look up and see the name on the bus. The name in the book and on the bus are related, but they also have meaning for me. If I am serious about redressing the unintended insult, seeing the name on the bus after seeing it in the book enforces my decision to do this. And if I happen to be what many people would call superstitious, I may feel that seeing the name on the bus after reading it was a nudge to do just that. The universe, I might say, is telling me to take care of this. At this point, the coincidence has become meaningful; it seems to be *communicating* something to me.

In this case, our hypothetical coincidence itself was not that remarkable, but the meaning it seemed to convey was significant enough to cause a pang of conscience and the determination to rectify a mistake. In some cases we will look at, the coincidence itself seems incredible, yet whatever meaning it may have carried seems negligible if not absent. With these coincidences, one is tempted to say that the medium is the message, that the coincidence itself, its utter improbability, is the meaning to be conveyed. As does precognition, such coincidences suggest that the world might not be quite the way we think it is. It might be rather different.

Kammerer and coincidence

In the 1920s, the Austrian biologist Paul Kammerer, another 'collector of coincidences', developed a 'classification of coincidences', according to the type, the number of coincidental elements involved, and other factors. Here I will look at some coincidences, ranking them according to how close they come to breaking the 'believability barrier'. I'll start with a few of my own.

A few years ago, I met with some friends to have dinner at an Indian restaurant here in London. When the bill came I laughed: it was for £123.45. More recently, while writing this book, I took a break and looked at the word count for the chapter I was working on. When I saw that it was 6666 I laughed, and posted a photograph of it on Twitter.[4] In the first case, the coincidence is clear: the bill came to the first five natural numbers. Nothing particularly mystical or special about that, it just doesn't happen very often. In the second case one has to know that 666 is the number of the Beast in Revelations, and that it is also associated with the notorious Aleister Crowley, devil worship, Satan, and so on. So the coincidence involves being aware of the connotations of 666, a bit more knowledge than that of the natural numbers.

Many years ago, when I had left Los Angeles at the start of my move to London, I stopped in New York for a few days. On the day I was supposed to leave, the blizzard of 1996 hit; all flights were cancelled and I was snowed in. When I went to the one open café near my hotel I overheard a conversation. At a table was someone named Gary; he was also on his way to London from Los Angeles, and was also snowed in, waiting for a flight.

More recently, while researching this book, I was happy to discover a second hand bookshop I didn't know of. While there I saw a stack of books with a card on top of them. On the card a name was written. I asked the bookseller what name was on the card. He told me it was Gary. I told him that was my name. He told me it was his. I mentioned that many years ago I worked at a second hand bookshop and that I was researching a book about coincidences.

Speaking of the bookshop where I used to work, more than a few coincidences happened there; I've already mentioned my precognitive dream about the Tibetan monks and the mandala. Once, while I was reading a book about colour theory, someone in the office answered a call. It was a customer looking for books about colour theory. On another occasion, I asked for the title of the 'new age' music we were playing. As soon as I was told that it was 'Alpine Blizzard', I saw that I was about to send a notification about a book to a customer named Snow.

Some years ago, I was on holiday in Ticino, the Italian-speaking part of Switzerland. On the way to the house in the hills that I had rented, the old Kinks song from the 1960s, 'Party Line', came to me, about an old telephone service that had more than one customer using the line. I hadn't heard it in years and nothing on the journey suggested it, but I couldn't get it out of my head. When we got to the house, I looked around and saw that the telephone was in a kind of booth, rather like an old pay phone. I picked up the receiver, but instead of a dial tone I heard a voice. It was a party line.

Once, I was sitting in the Hampstead Parish churchyard, not far from where I live, reading one of Sax Rohmer's Fu Manchu novels. As I read a passage in which a church bell strikes 7:00, the bell in Hampstead Parish church struck seven too.

Many years earlier, when I still lived in Los Angeles, I picked up the telephone to call F., my 'future ex-wife', who was at work. Before I could dial I heard a voice – the telephone hadn't rung. It was our marriage counsellor, returning a call from F. I gave him her work number, the one I was about to call.

Here are two more coincidences involving my ex-wife. One evening, F. told me that *Mad Love*, the Peter Lorre remake of *The Hands of Orlac*, was showing at the cinema (I've mentioned my fondness for old horror films). We were too late to catch it, so we rented *Under the Volcano*, the John Huston film of Malcolm Lowry's novel. In it there is a scene in which Geoffrey Firmin, played by Albert Finney, passes by a cinema. What is playing? *Mad Love*.[5] On another occasion, in the *Los Angeles Times*, F. read an article about Chiune Sugihara, the 'Japanese Schindler', who saved many Jewish lives while he was a diplomat in Lithuania in the early days of the Second World War.[6] (I've mentioned that she was half-Japanese.) She hadn't heard of him before. The next day, on impulse she decided to open a letter that was addressed to her mother but had been delivered to her by mistake. She had no idea what it was, and had been holding on to it for a year and had forgotten about it. When she opened it she discovered a letter and a clipping of an article about Chiune Sugihara.

Degrees of coincidence

How do these coincidences rank? The first two score on more than one coincidental element. Both have two people named Gary; in fact, in the second there are three: myself, the bookseller, and the 'Gary' whose name was on the stack of books. (I never found out who that was and I can't say why I asked about the name in the first place.) In the first, the two Garys were both on their way to London from Los Angeles, had both been snowed in and so missed their flights, and had both come to the same café at the same time, which suggests they were both staying in the same neighbourhood. In the second, the two Garys share a background in being booksellers, and this coincidence comes to light while one of them is researching a book about coincidences. The two coincidences from my time at the bookshop are the sort that raised a laugh and some eyebrows and had my mystical-minded workmates nodding their heads. The church bell ringing at the same time that I read of one doing the same, was an example of what I call the 'guardian angel' effect. I was working on my book about Madame Blavatsky at the time and took it as a sign of approval; Sax Rohmer had even mentioned Blavatsky in the novel. The 'Party Line' coincidence may be precognitive. I somehow knew in advance that the telephone at the holiday house would be on a party line. Or perhaps I should say that my right brain knew this. How did it communicate this knowledge? Through a song, that is, music, which is something the right brain is very partial to. But why it was important for me to know, I can't imagine. Mobile phones were common by then, and I only picked up the receiver out of curiosity.

With the coincidences involving my ex-wife, more factors come into play. With the telephone call from the marriage counsellor, three people were involved: the counsellor, my ex-wife, and myself. He wanted to reach her at the same time that I did, and the number he needed to call was the one I was about to dial. That it was our marriage counsellor suggests emotions were high and the atmosphere tense, conditions frequently associated with telepathy and other paranormal phenomena.

The *Mad Love/Under the Volcano* connection involves my ex-wife, myself, the decision to see a film, missing that film, but seeing it at least mentioned in another film, which is based on a novel in which the film we missed plays an important part. Both films revolve around doomed loves, Firmin's (Albert Finney) for his ex-wife, and Dr Gogol's (Peter Lorre) for a woman who rejects him. Lowry was fascinated with coincidences, signs, portents, and was a reader of occult and esoteric literature.[7] Ouspensky's ideas about recurrence and the Tarot run throughout the novel, as do Dunne's ideas about time; they are less prominent in the film. And the coincidence involving Chiune Sugihara is even more complicated. It involved a letter written to my ex-wife's mother but delivered to her by mistake, which contained a clipping about Sugihara, which she didn't open until a year after it arrived, and only on the day *after* she had read the *Los Angeles Times* article about Sugihara. Until then she knew nothing about him.

Plum pudding

One of the best-known series of coincidences, which turns up in most books on the subject, concerns a Monsieur de Fontgibu and a plum pudding. When he was a schoolboy Emile Deschamps, a

French poet, was given a piece of plum pudding, an English delicacy, by a Monsieur de Fontgibu. It was a rare dish in France at the time. Ten years later Deschamps saw some plum pudding on display at a Paris restaurant. When he asked if he could have it, he was told it was the last one. Someone had already ordered it. Who? Monsieur de Fontgibu, who offered to share it with him. Years later, Deschamps was at a party where plum pudding would be served. He told the story of Monsieur de Fontgibu, and said all he needed now was for him to turn up. As they sat eating the plum pudding, the servant opened the door and announced Monsieur de Fontgibu, who by this time was an elderly gentleman. He had been invited to a party at another apartment in the building, and had come to the wrong door.

The French scientist Camille Flammarion, who tells this story, was himself involved in a remarkable coincidence. He was working on a book when a gust of wind blew his pages out of the window. As it did, it began to rain. Flammarion thought it would be futile trying to retrieve the pages and let them go. When, a few days later, the chapter he had been working on arrived from his printer, he was amazed. The pages he had lost were there. What had happened was that the porter from his printing house was walking past Flammarion's window. He saw the pages on the ground and assumed he had dropped them himself, so he gathered them together and took them to the printer. And what had Flammarion been writing about? The wind. No wonder that it was Flammarion who coined the term 'collector of coincidences'.

The library angel

There is a peculiarly literary kind of coincidence that is the work of an agency that Arthur Koestler has christened 'the library angel'. These are coincidences in which material a writer or researcher is seeking seems to arrive at hand as if by magic, a phenomenon I have benefited from more than once. In *The Challenge of Chance*, written with Sir Alister Hardy, a scientist interested in parapsychology, and Robert Harvie, a psychologist trained in statistical method, Koestler offers some examples.

This one comes from Koestler himself. In 1972, Koestler was asked to write about the world champion chess match between Boris Spassky and Bobby Fischer, which was to be held in Reykjavik, Iceland. Koestler was a lifelong chess player, but he wanted to catch up on any recent developments in the game. He also wanted to brush up on his knowledge of Iceland, which was minimal at best. He went to the London Library to borrow some books on these subjects, and unsure where to start, he decided to head to the section on chess. Of the thirty or so books on the shelves the first one to catch his eye was *Chess in Iceland and Icelandic Literature* by Willard Fiske. It was as if it had been waiting for him.

Some material is even more handy. The novelist and journalist Rebecca West was researching some of the records of the Nuremberg trials. At the Royal Institute of International Affairs, she discovered that the trials were arranged in abstracts under arbitrary headings, making it almost impossible for a researcher to locate anything. After hours of fruitless searching, she told the librarian that she was unable to find what she was looking for. Staring at the shelves of volumes, she said, 'It could be in any of these,' and to make her point, she pulled a volume down and opened it to illustrate. She was surprised to discover that not only had she pulled out the correct volume, she had opened it to the exact page she needed.

I can't resist including another example from Rebecca West, not so much about the library angel providing what was needed, but literary nonetheless. She writes that 'Once in the South of France I was writing a passage about a girl finding a hedgehog in her garden, when a servant told me to come out and see a hedgehog they had found in the garden.'[8] As a hedgehog began this book, I thought it fitting to include another here.

Melchizedeks needed?

The library angel even works outside of the library. Jacques Vallee, the ufologist, (the film director Francois Truffaut plays him in the film *Close Encounters of the Third Kind*) tells a story of when he was researching an obscure cult that used the name of Melchizedek, a Biblical figure. He spent a great deal of time researching every reference he could find. When he asked a Los Angeles taxi driver for a receipt, he was surprised to see that her name was M. Melchizedek. Assuming there must be a few Melchizedeks in town, when he checked the enormous Los Angeles phone directory, he was even more surprised to discover that she was the only one listed. It was as if he had sent his request out into the ether and the library angel said 'How about this one?'

Colin Wilson, who writes about this coincidence, had his own Melchizedek experience. Wilson was writing a chapter on synchronicity for *The Encyclopaedia of Unsolved Mysteries*, and, as coincidence would have it, a number of coincidences gathered around him as he did. For example, another article for the encyclopaedia concerned Joan of Arc. Wilson had a bound series of the *International History Magazine*, and decided to see if it could provide any help. He opened it at random to an article about Joan of Arc. Shortly after he received a copy of a biography of Ayn Rand by her ex-disciple Barbara Branden. He was reading this when the post arrived. In it was a letter from a reader who remarked that Barbara Branden mentioned him in her recent biography of Ayn Rand.

Other coincidences followed, but what topped them off was another Melchizedek. Breaking off from writing his chapter on synchronicity, Wilson noticed a book in a pile strewn on his study floor. It was titled *You Are Sentenced to Life*; the author was W. D. Chesney and it was privately printed in California. Wilson didn't remember buying it and he had certainly not read it. He paged through it and at the very end he read 'Order of Melchizedek', the heading of a letter from an 'Instructor within the Order of Melchizedek' reprinted in the book. As Wilson writes, 'I doubt whether, in the two thousand or so books in my study, there is another reference to Melchizedek; but I had to stumble upon this one after writing about Melchizedek in an article about synchronicity.'[9]

Fragrant Floriline

Koestler himself was inundated with remarkable coincidences when he was writing about them. It is as if the very act of thinking about coincidences attracts them, an example of the 'clustering' effect that Paul Kammerer, a collector of coincidences, argued was a kind of universal law, operating on the ancient Hermetic principle that 'like attracts like'. We will return to Kammerer but before we do, here are two more coincidences that seem too improbable to be true.

William Morris, leader of the Arts and Craft movement, had purchased supplies for a trip to Iceland – and is this a coincidence, after Koestler's chess match? Shortly before leaving, he heard from the supplier that they had packed someone else's order in with his. Morris asked them to come and open the packing case and retrieve their items, otherwise they would be on their way to Iceland. The supplier never came, so the case went with Morris and his companions. The supplier never mentioned what the mistaken item was, and during the journey Morris and his companions guessed what it could be. 'As the wildest possible idea', Morris suggested that it was 'Fragrant Floriline' – a kind of tooth paste popular at the time – and hairbrushes. When it came time to open the case, the group was curious to see what the mysterious item really was. What was it? Four boxes of Fragrant Floriline as well as two dozen bottles of flower scents, packed in boxes with hair-brushes printed on them. Understandably, Morris and his group didn't know if they were 'drunk or dreaming' and they 'rolled about and roared'.[10]

While on a journey through Manchuria to China, the explorer Peter Fleming, brother of Ian Fleming of James Bond fame, told his companion the story of his part in the expedition to find Colonel Fawcett, the early twentieth century explorer who went missing trying to discover the 'lost city of Z' in the jungles of Brazil; his story was recently made into a lacklustre film. While telling the tale, Fleming remarked on Fawcett's 'immortality', and how others had never given up on him. As he did, his companion saw a scrap of an English newspaper on the otherwise unblemished ground. When he picked it up he saw that it was part of an article about a recent expedition to find Colonel Fawcett, and, as they had just done, it too mentioned Fawcett's 'immortality'. What a scrap of English newsprint was doing on a virgin Manchurian landscape is anybody's guess. But that it should report on the very incident Fleming was relating beggars belief.

One last coincidence, which must surely have been the work of an angel. An architect suffering a nervous breakdown decided to end it all by throwing himself in front of a train. As he did a passenger on the train, completely unaware of the architect's anguish, gave in to a sudden impulse to pull the emergency brake. The train stopped just before reaching the architect. The unrelated impulse to behave anti-socially saved his life.[11]

The ink-fish effect

Koestler writes that some coincidences lead one to suspect that otherwise trivial events are part of a plan, a design, 'related to that striving towards higher forms of order and unity-in-variety', which he calls the 'integrative' or 'self-transcending' tendency in the human psyche and the universe itself.[12] Yet others, which are 'merely an insult to the laws of probability' are subject to what he calls the 'ink-fish effect', a kind of deliberate clouding of the issue so that we can't arrive at any acceptable account of them.[13] Some coincidences seem 'purposefully arranged', while others are 'just impish or whimsical'.[14] How do we tell the difference?

Unless we discover some so far unknown meaning in Emile Deschamps' preposterous encounters with Monsieur de Fontgibu and the plum pudding, or in William Morris scoring a bull's eye with his 'wildest possible idea' that Fragrant Floriline would be in the packing case, or in Peter Fleming's companion finding an article on Colonel Fawcett in a scrap of newspaper in Manchuria just as they were speaking of him, I suggest that we can collect these coincidences under 'the sign of the ink-fish', as Koestler puts it. Aside from their sheer improbability, there doesn't seem much meaning to be had in these bizarre examples. Our examples of the library angel at work are a different matter. In them someone was actively looking for information that the angel promptly provided. Deschamps may have wanted his plum pudding, but he didn't necessarily want to see Monsieur de Fontgibu. And there is no evidence that William Morris wanted Fragrant Floriline. So perhaps we can say that when human desire or intention is involved, the ink-fish departs and something more constructive begins to operate.

In *The Roots of Coincidence* Arthur Koestler suggests that so far only two theories have made an attempt to come to grips with 'meaningful coincidence', Jung's synchronicity and Paul Kammerer's 'seriality', which is rather different than Dunne's Serialism. Of the two, Kammerer's seriality is less known.

Lamarckian heresy

In *The Case of the Midwife Toad,* Koestler tells the story of Kammerer's suicide, when it appeared that the results of his experiments proving the inheritance of acquired characteristics had been falsified. Most likely Kammerer himself did not fake his results; there is good reason to believe that the falsification was the work of either a loyal but misguided assistant, or that of a rival out to discredit him. In either case, it did not disprove Kammerer's research, merely provided his critics with a reason to ignore it. The 'inheritance of acquired characteristics' is the idea that the achievements of adults can be passed on to their offspring through their genes. It is associated with the eighteenth century French naturalist Jean-Baptiste Lamarck. So, for Lamarck, giraffes acquired their long necks because earlier generations of giraffes stretched to reach leaves higher on the trees, and the results of their efforts were passed on to their offspring. The 'Lamarckian heresy' is anathema to strict neo-Darwinians because it argues that effort, that is, desire and action, can effect evolution. For them this is nonsense; evolution is the product of random mutations and the 'survival of the fittest'. Will and desire have nothing to do with it. Lamarck thought differently.

Sadly we can't debate the question here, although there is plenty of evidence for the inheritance of acquired characteristics, going

back to Darwin's first philosophical – not religious – critic, Samuel
Butler.[15] But another interest of Kammerer's was coincidence,
and, as mentioned, he developed a theory that argued that similar
things happen or gather together. He called it 'seriality' which
means that similar things happen in a series. There was nothing
occult or paranormal about Kammerer's idea; he saw it as a kind
of 'law' or force, like magnetism, that somehow drew like things
together. My examples of coming upon another Gary who, like
me, was snowbound in New York while *en route* from Los Angeles
to London, or of meeting a Gary in a bookstore who, like me, was
a bookseller, while I was researching material on coincidence, are
expressions of what Kammerer would have called a 'cluster'.

Seriality

Kammerer wrote a book explaining his theory, *Das Gesetz der Serie*,
which roughly translates as 'the law of seriality'. Einstein thought
it was 'original and by no means absurd'. It could be understood as
an attempt to prove the adage that it 'never rains but it pours', or
to supply sound reasons for the popular belief that 'things come in
threes', or that celebrities seem to die in threes.[16] Kammerer kept a
log book of coincidences for more than twenty years and his book is
filled with examples. So he tells us that on two successive days, the
number of his brother-in-law's seat at a concert and on his cloakroom
ticket were the same: 9 on the first day, 21 on the second. While in
her doctor's waiting room, Kammerer's wife read an article about an
artist named Schwalbach; as she made a mental note to remember his
name, the receptionist called for Frau Schwalbach. Another example
concerned the name 'Rohan', which turned up in a novel his wife was
reading. She then saw someone on a tram who looked like her friend,
also named Rohan; later that evening the real Rohan dropped by their
apartment, unannounced. In the tram, she had heard someone ask
the Rohan lookalike if Weissenbach, a village on Lake Attersee, would
be a nice place for a holiday. At a delicatessen Kammerer's wife went
to after leaving the tram, the owner asked if she knew of Weissenbach;
he had to post a parcel there but did not know the postcode.

The first coincidence was of the 'second order', according to Kammerer, because it involved the same type of coincidence happening on two successive days. The other was of a 'progressive type', as the coincidences proceeding from one to the other, from Rohan to Weissenbach.

Kammerer understood a series to be 'a lawful recurrence of the same or similar things and events ... whereby the individual members in the sequence ... are not connected by the same active cause.'[17] Which means that they are brought together by something other than cause and effect. What this something other is, for Kammerer, is seriality. This is a 'universal principle in nature', operating '*independently from physical causation*' (Koestler's italics).[18] For Kammerer, what we experience as odd coincidences are merely the few manifestations of seriality that catch our attention. But its effects are in operation all the time (rather as we rarely notice the future turning up in our dreams, although Dunne says it's there all the time). For Kammerer, the laws of seriality are as ubiquitous as those of physics; we simply fail to notice them.

Cosmic kaleidoscope

But Kammerer did. Seriality worked like gravity, but where gravity pulls everything down to earth, seriality is more selective and fastidious, more discriminating. It pulls similar things together. It was a 'cosmic kaleidoscope', which, no matter how it was turned, always brought 'like and like together'.[19] In a way it presented a correspondence among different phenomena in the way that T. C. Lethbridge believed his experiments with the pendulum had revealed. In space it brought things together simultaneously: similar things finding themselves all in the same space at the same time with no adhesive other than their similarity. In time it arranged them in sequence: the same sort of thing following one after the other, again with nothing but their similarity linking them. Kammerer made notes of the people he saw as he sat in Vienna's parks, observing them; he made notes of the people he saw as he rode the trams, what they wore, what they talked about, where they got on and where they got off. He came to the conclusion that 'the recurrence of identical or similar data in contiguous areas of space and time is a simple empirical fact.' This led to the recognition that coincidence, rather than the rare exception, is the rule 'to such an extent that the concept of coincidence itself is negated.'[20] When everything is a coincidence, nothing is. It is only our ignorance of the law of seriality that gives us the impression that coincidences are anomalies.

Law of large numbers

In some ways, Kammerer's seriality is similar to what in probability theory is known as the law of large numbers. This refers to the strange fact that if enough random events are collated, they will produce stable, ordered, repeatable effects; the insurance business is based on this inexplicable fact. A classic example is the number of people bitten by dogs in a single day. According to statistics provided by the New York Department of Health, this proves to be a remarkably stable number: in 1955 it was 75.3; in 1956, 73.6; in 1957, 73.5, and so on. No one can tell which dog will bite which person – that's the random part – and how the dogs themselves know when enough people have been bitten so they can stop, isn't clear, but the statistical average remains constant. I haven't seen statistics for more recent years but I imagine they are not radically different. Other examples are the number of murders in a given year; we don't know who will be killed or by whom, but the average is known and remains stable. Likewise suicides. Like Kammerer's seriality and the more recent and related chaos and complexity theory, the law of large numbers is an example of a *large number of uncertainties producing a certainty*, of 'random events creating a lawful outcome'.[21] There is no physical explanation for this, in terms of some force or energy making it so. No one quite knows why it happens. As with seriality, it seems that the world 'just is like that'.

Cycles

Kammerer was interested in cycles. The 'lucky streaks' that come to gamblers are a product of temporal cycles, as are the good and bad days we all experience. In a sense Kammerer was charting the kind of differences in time that the Hopi and P. D. Ouspensky were aware of, how some days are propitious while on others we should stay in bed. If the Egyptians who were aware of the different *neters* responsible for the quality of a day were aware of Kammerer's work, they would have appreciated it. In the book I wrote before this one, *The Return of Holy Russia*, I discuss the ideas of some Russian scientists who believed human life was conditioned by cosmic forces, sunspot cycles, for example, or cosmic rays.[22] It is not quite astrology, but it is in the same ballpark. They might not agree with Kammerer about seriality, but like him they rejected the notion of time as bare temporal extension. It has a character and Kammerer believed it could be charted. For Kammerer, seriality was 'the umbilical cord that connects thought, feeling, science and art with the womb of the universe which gave birth to them.'[23] It is a shame his suicide did not allow him to develop his ideas. Scientists today could be testing for seriality in the way they do for precognition.

Jung, the *I Ching* and synchronicity

The other candidate for a theory of the paranormal that takes on the challenge of meaningful coincidence, at least according to Koestler, is Jung's synchronicity. This is well known, or at least the word 'synchronicity', coined by Jung, is. Along with 'complex', 'archetype', 'introvert', and 'extrovert', it is one of the Jungianisms that have made their way into common discourse. In many circles, and not only 'new age' ones, it has replaced the more pedestrian 'coincidence'. 'What a synchronicity,' many people say today when what they mean is 'What a coincidence.' Do they understand what synchronicity means? Not always. But then Jung, who thought of it and wrote a book trying to explain it, wasn't that sure himself.

Koestler has strong words for Jung, who was not the best

expositor of his own ideas. He respects his insights, but balks at his presentation. As I point out in my book on Jung, one reason for this was Jung's insistence on presenting his ideas in a 'scientific' manner, which hamstrings his major works with a verbose 'Herr Doctor Professor' prose whose overall effect can be numbing rather than enlightening.[24] Which is not to say that the ideas are not important and enlightening. They are. It's just that Jung is well served by writers like Marie-Louise von Franz and Anthony Storr – two Jungians widely apart in their approach – who can distil from Jung's bubbling alchemical alembic the essence of his vision. Jung too, when he is 'unbuttoned', as he is on occasion, can be forthright and straightforward. It's just that this is a rarity.

In *Synchronicity: An Acausal Connecting Principle*, part of a volume co-authored with the physicist Wolfgang Pauli, Jung presents synchronicity in some of his most tortuous prose, decked out with charts, statistics, and other 'scientific' impedimenta. But in the foreword he wrote to Richard Wilhelm's translation of the *I Ching, or Book of Changes*, an ancient Chinese oracle and book of wisdom, which was published earlier, he was less obtuse. Jung had been using the *I Ching* for years, but it wasn't until late in life, following a near death experience, that he came out of the closet, as it were, if not as an occultist, than certainly as someone with a deep interest in a variety of occult ideas. 'Synchronous' means 'occurring at the same time', and this is what synchronicity is supposed to refer to: 'the coincidence of events in space and time' that mean 'something more than mere chance'.

So far this is what a 'meaningful coincidence' is, so one could be excused for wondering why Jung felt the need to invent a new word for it. The coincidence, however, is not only among events – it could then fall under Kammerer's seriality, about which Jung knew – but also between events and 'the subjective (psychic) states of the observer or observers.'[25] So the real coincidence is between what is happening in the outer world, and what is happening in my inner one, what Michael Shallis calls an 'internal-external mirroring'.[26] This mirroring happens at the same time – what mirroring doesn't? – hence the events mirrored are synchronous. The *I Ching* is a book of synchronicity, or at least it provides one half of the mirroring needed for it.

A toss of the coins

For readers unfamiliar with the *I Ching* I should explain that it is a method of inquiring into the character and quality of the time at the moment one asks a question. It is related to the Taoist notion of the two principles at work in the world, yin, or the dark, and yang, or the light, mentioned in Chapter Three in association with our cerebral hemispheres. The traditional method used yarrow stalks and a tedious process of elimination. Today most people use the coin method. Take three of the same coins, give heads a value of 3 and tails a value of 2. After focusing your question, toss the coins six times. Odd values (7 or 9) give a straight line (—), even values (6 or 8) give a broken line (– –). These lines are stacked one on top of another, until a hexagram is formed. Nines and sixes produce what is called a 'moving' line. This becomes its opposite and so produces a second, additional hexagram (one repeats the other, non-moving lines to form it). There are sixty-four hexagrams, each one providing an answer to your question, and a commentary on whether the time bodes well for whatever it is you are asking about.[27] In a sense the hexagrams making up the *I Ching* are similar to Jung's archetypes in that they are supposed to represent common themes and situations in life, something the Tarot does too.

How does it work? That's a good question.

The first four hexgrams of the I Ching

Take a chance

'The Chinese mind,' Jung wrote, 'seems to be exclusively preoccupied with the chance aspect of events.' Coincidence concerns it much more than causality. 'The moment under actual observation appears to the ancient Chinese view more of a chance hit than a clearly defined result of concurring causal chain processes.' The Western mind (read, left brain) selects, discriminates and calculates, but for the devotees of the *I Ching* what is important is 'everything down to the minutest nonsensical detail', because all of it makes up the 'observed moment' (right brain, no?).

Earlier in this chapter I said that as I sat here, at my desk, working, there were many other things happening, but that we could not consider them part of a coincidence, because they did not seem related to me. What Jung is saying is that all of those things, including myself and the *I Ching* coins I have just thrown, are part of the same moment, and because they are, they all share in the quality of that moment. We can see that the *I Ching*, like the Hopi, is aware of time's character. What appears to happen when consulting the *I Ching* is that my question and the coins distil the essence of the moment, whose character is captured in the hexagram my chance toss provides. Out of the simultaneous 'happenings' that constitute reality at that moment, my question and the *I Ching's* response are a kind of snapshot, 'capturing the moment'.

Theoretically, anything happening at that moment shares in that moment's character, so conceivably anything could work as an oracle. This is why tea leaves and other more pungent materials – such as animal entrails – have been used as a means of 'seeing the future'. The story around the origin of the *I Ching* was that it began with reading the cracks in a tortoise shell. And while I have not read tea leaves, nor any poor animal's viscera, I have used the *I Ching* for as long as I have recorded my dreams and I can say that it works.

In his foreword, Jung says that the hexagram is 'an indicator of the essential situation' prevailing at the moment, and it operates according to a 'curious principle' he has termed synchronicity. Where causality tells us how A led to B which led to C, synchronicity is concerned with how A, B, and C all turned up the same moment, why they form a cluster, as Kammerer would have said. Part of that cluster is what is in your mind at the moment of tossing the coins. But unlike precognition, we cannot test for synchronicity, just as we can't test for coincidences. And if an essential ingredient in a synchronicity is the subjective state of the person asking the question, we can't test for that either. The moment is unique and cannot be repeated. If you tossed the coins again, trying to arrive at the same result, the *I Ching* itself will tell you that is cheating.

The living soul of the book

Hexagram No. 4, *Mêng*, is called Youthful Folly: 'It is not I who seek the young fool;' it says, 'the young fool seeks me. At the first oracle I inform him. If he asks two or three times, it is importunity.'[28] As someone who has consulted the *I Ching* and been unhappy with the result and so tossed the coins again, asking the same question, and then received this reply, I can attest to the shock of having an inanimate text give as direct an answer as could be imagined. But is it really an 'inanimate text'? Jung himself wasn't sure.

Jung admitted that 'the only criterion of the validity of synchronicity is the observer's opinion that the text of the hexagram amounts to a true rendering of his psychic condition.'[29] He relates this to the knack a wine connoisseur has for detecting the vineyard and vintage of a particular wine, or that of an antique dealer for placing an *object d'art* in its period and provenance, or even an astrologer's ability to tell your sign merely by meeting you, all of which seem examples of the kind of 'implicit' knowing discussed in Chapter Three, what the seventeenth century religious thinker and logician Blaise Pascal called the 'spirit of finesse', in distinction to the 'spirit of geometry'. There is no test for this, just as there is no test to determine whether your or my or a psychotherapist's interpretation of a dream is correct. Or, rather, there is a test and it is called 'life'. Just as the proof of the pudding is in the eating, the proof of the *I Ching's* accuracy is in how the reading affects you and your actions. If a hexagram 'clicks' you will know soon enough. I have had enough experience of the *I Ching* to know that it is never wrong. On the other hand, I am as guilty as anyone of not accepting a reading and importuning the oracle to provide another more in line with my desires. In doing so, on more than one occasion I received *Mêng* and went away, temporarily chastened.

It is this character of intelligence – like that often displayed in dreams – that led Jung, in an unbuttoned moment, to speak of the 'living soul of the book' and to remark that the traditional explanation for how the oracle works, is that 'spiritual agencies' guide the coins so that they provide an appropriate answer. We may not want to follow Jung here, but I see no categorical reason not to. How else can I explain how a book and a few coins could *know* that I am asking the same question twice? For this is the effect of having the *I Ching* say 'I answered you already. Go away and don't bother me.' And when it has done that often enough, spiritual agencies no longer seem so far-fetched.

The doctor is in

But in Jung's major work on synchronicity, spiritual agencies no longer make an appearance. Bad enough that Jung was writing about an oracle; one assumes that prudence suggested he couch his exposition in his best (or worst) Herr Doctor Professor mode. The problem here is that this leads to more confusion than clarity. As he often does with his notion of the archetypes, synchronicity becomes a kind of all-purpose term that Jung applies to a variety of paranormal phenomena, including precognitive dreams. Yet, if synchronicities involve events and psychic states happening simultaneously – which, we remember, is what synchronous means – then precognitive dreams are a phenomenon of a different colour. Some time passes between the dream and the reality it knew of in advance, even if it is only minutes, so they are not synchronous. (Jung gets over this hurdle by saying that in the unconscious, time and space are relative, and so future and past are neighbours.) And, as we've seen, most precognitive dreams are about trivial events – aside, of course, for those that foresee a disaster – and are not particularly meaningful in the way that synchronicities are. The two are related, but not the same.

And this is true also of the results of J. B. Rhine's experiments with ESP, which Jung tries to corral under synchronicity. Jung also says that synchronicities occur when an archetype is 'activated'. I may be dense – assuredly I am – but I am not sure what that means, other than a state of emotional arousal and excitement.

But if this is what he means, why not say that? Jung also brings in statistics, astrology – of which he was a devotee – and quantum physics in order to support his notion of synchronicity as an 'acausal connecting principle', which, as Koestler points out, is something of an oxymoron. What Jung means is that, as with seriality, the 'cause' bringing the synchronistic elements together is not that of 'cause and effect'. But it is also something other than chance, because it is meaningful. Chance does not supply meaning; that's why people who reject synchronicities say they are 'only coincidences' and mean nothing. So in the end it seems that synchronicities are meaningful coincidences that happen for no reason, which somehow can't be right.

Out of all the examples Jung provides, really only one is a synchronicity, in the true sense. It is mentioned in practically every book on synchronicity and on Jung. One of his patients was an infuriatingly rationalistic woman whose staunch intellectualism made therapy tough going. As she told Jung of a dream she had about a golden scarab, Jung heard a knocking at his window. He opened it and a golden-green scarab beetle flew in. They were rare to begin with and not taken to knocking on windows. Jung's patient was stunned at the coincidence; her rationalism cracked, and Jung was able to get on with his therapy.

Jung once mentioned another, perhaps even more apt example, to Esther Harding, a leading Jungian. A woman he was treating refused to see any expression of sex in her dreams as anything but symbolic – a neat turn around, as this was often his tack when arguing with Freud. Jung felt she needed to accept sex in its physical sense, something she resisted. Jung made little headway until at some point 'two sparrows fluttered to the ground at her feet and "performed the act".'[30] His patient got the point after that. This may be an even better example, as it could be said that Jung's rationalist patient had a precognitive dream about the scarab, although it isn't known exactly how the scarab in her dream corresponded with the one in Jung's consulting room – did she dream of seeing the scarab there, or just of a scarab? And the scarab did fly in as she was telling Jung about it, so that was synchronous.

Transcendent functions in the *Unus Mundus*

But what is important in both examples is that the synchronicity had a profound impact on the people involved; it meant something to them and was an agent in their psychological development, their individuation. They were examples of what Jung called the 'transcendent function', which usually appears in dreams. It is when a dream provides a symbol that miraculously unites opposites between which the dreamer has been caught in waking life in an unbearable tension, stuck, as we say, 'between a rock and a hard place'. The conscious, rational mind sees no solution and is at its wits' end. The dream provides a symbol that 'transcends' the opposition, raising the psyche 'above' the tension and seemingly irresolvable conflict. One's awareness is broadened and one finds that the conflict is no longer a problem, not because it is solved, but because it has been outgrown.

As Kammerer did with seriality, Jung wanted to find some way to conceive of synchronicity as a kind of universal principle, and he suggested that his exploration of the archetypes of the collective unconscious were a psychological parallel with the exploration of sub-atomic particles. Both were part of what he called the *Unus Mundus*, or 'One World', an underlying unity from out of which what we experience as our inner and outer worlds emerge. Archetypes for Jung had what he called a 'psychoid' character, a, for us, inconceivable state of being that is neither physical nor mental but the unmanifest ground of both. In this sense it is similar to what the physicist David Bohm called the 'implicate order', a state of 'unbroken wholeness' that underlies our phenomenal world. Note that this order is 'implicate', that is, implicit, not explicit, it cannot be pinned down in any detail. As Marie-Louise von Franz points out, this is a view embraced by the *I Ching*: it 'presumes that the whole of nature is a psychophysical unity, or has a unitary wholeness, which, however, escapes observation that concentrates on details.'[31] Or, as J. B. Priestley put it, 'sharp analysis and precision may trap us in a dead end,' anticipating McGilchrist's arguments about the danger of a too dominant left brain.[32]

The Pauli effect and the synchronistic frisson

The physicist Wolfgang Pauli, who worked with Jung, agreed with this; his contribution to their joint work concerned the influence of archetypes on the work of the astronomer and astrologer Johannes Kepler, with Copernicus and Newton, one of the Three Musketeers of pre-Einstein astronomy. Earlier I pointed out that Pauli himself had a knack for what we might call 'synchronistic disasters' or 'disruptive coincidence'. There are scores of stories of Pauli's presence precipitating some catastrophe in the laboratory; it came to be known as 'the Pauli effect'. There is even an account when a disaster happened, and someone remarked that at least on this occasion, Pauli couldn't be held responsible as he wasn't there. It later turned out that he was in a nearby train station when the catastrophe occurred. One of Pauli's major contributions to physics is what is known as the 'Pauli Exclusion Principle', which states that no two electrons in an atom can have the same quantum numbers. Friends decided to extend this to Pauli himself, and said that Pauli and working laboratory equipment cannot occupy the same space. Or going by the train station story, even the same neighbourhood.

Readers may take me to task for this, but I have to say that Jung's attempts to account for synchronicity in terms of what he calls 'acausal orderedness', which he associates with natural numbers and other 'just so' characteristics of the physical world are, to my mind, his least convincing arguments. And this is not only because the Herr Doctor Professor effect is unrelievedly in evidence in them. This tack had been taken up by his eloquent disciple, Marie-Louise von Franz, in a number of essays, and has even been the subject of a book by F. David Peat, a colleague of Bohm, in which he tries to establish synchronicity as the lynch pin holding together coincidence, consciousness, and practically everything else.[33] Their work and that of others along this line is, of course, important. But I have to say that personally, this side of the synchronicity debate leaves me dissatisfied. I simply don't see what quantum physics has to do with the peculiar *frisson* I experience when the synchronicity bell rings – as it did when, on my way to lecture about Colin Wilson's 'outsiders', I saw a copy of *Vogue* with all of the print on it obscured, except for 'The Outsider'. I am not saying sub-atomic

particles weren't involved. I just don't see how I get from positrons
to the unmistakable shiver I feel when this happens.

What is the essence of this feeling? That somehow, something
– someone? – *knows* what I am thinking or feeling or am on my
way to do at that moment and lets me know that it knows in a way
seemingly designed to make me get the hint – as Jung's prudish
client did, when she saw two sparrows 'perform the act'. Whatever
it is, it knows this in the same way that the *I Ching* knows when I
am importuning it. And I guess it knows it in the same way that one
particle knows what the other particle it was just separated from is
doing although they are no longer in contact, what is known as
'non-locality' and 'quantum entanglement'. Or it knows it in the
way a wavicle knows when to act like a particle or to behave like a
wave. As I write this I wonder: are particles subject to synchronicity
too, rather than in some way being responsible for it? Maybe they
themselves wonder how all this weirdness is possible?

The participating mind

This does not mean that we must reject Jung's idea of an *Unus
Mundus*. In everyday terms, this means that, in some way we don't
quite understand, our inner and outer worlds participate in each
other, which, as we've seen, is the essence of a synchronicity. What
we want to know is the nature of this participation. I have explored
this in other books. What Jung seems to want to do is acknowledge
the participation, but avoid any active character to it, at least on the
human side of it. He is looking for a 'formal factor', to account for
synchronicity, 'which has nothing to do with brain activity.'[34]

Why is he looking for this? Because he wants to affirm the idea
that consciousness or the mind or the psyche is not dependent on
the brain. Unfortunately, he doesn't make this clear, and what he
does say leaves him open to some of Koestler's criticisms.[35] We can
appreciate Jung's desire to avoid reducing the psyche to the activity
of the brain, but to say that we must 'completely give up the idea
of the psyche's being somehow connected with the brain', as he
does, is overstating the case. I can say that the television programs

I watch are not reducible to the television, without saying there is no connection between them. (This is the 'receiver' picture of the brain: it doesn't generate consciousness, but 'tunes in' to it.) Dunne, Priestley, and Lethbridge all argue that mind and brain are not the same, but they do not say they are not connected. Consciousness and the psyche are connected to the brain, but are not reducible to it, just as the book you are reading is connected to the pages and ink it is made of, but is not reducible to them. The question is: what exactly is that connection?

Magic?

Jung, like some of his patients, found himself between his own rock and a hard place. He felt that assuming a 'causal relation between psyche and physis' – mind and body – led to 'conclusions … difficult to square with experience.'[36] What were these conclusions? Either that 'there are physical processes which cause psychic happenings, or there is a pre-existent psyche which organises matter.'[37] The first is crude reductionism: brain = psyche. The other is magic.

Jung abhorred being called a mystic or an occultist – my book *Jung the Mystic* looks at just this question – so any suggestion that we can make synchronicities happen, was for him a non-starter. They happen to us. But we've seen that Wolfgang Pauli apparently did make them happen – either that, or his appearance whenever a test tube burst or cyclotron broke down is one of the most remarkable series of coincidences on record. Jung himself scared the wits out of Freud when they were arguing about the paranormal and Freud's intransigent dismissal roused Jung's ire and led to the famous poltergeist in Freud's bookcase. In the middle of the argument a loud bang sounded from Freud's bookcase. When Jung said it was an example of what he called a 'catalytic exteriorisation phenomenon' – Herr Doctor Professor speak for poltergeist – Freud was even more dismissive. Jung disagreed and predicted that another bang would sound. On cue, one did and Freud was shaken. Jung, he believed, had somehow made it happen. There are accounts of the pots and pans at Jung's Bollingen Tower rattling whenever he brooded on an idea. Jung's psyche seemed to participate with the external world quite a bit.

When Rebecca West reached up and grabbed the right volume of Nuremberg trial abstracts and opened it to the right page – did she somehow make this happen? When I saw the copy of *Vogue* with 'The Outsider' on the cover while on my way to talk about *The Outsider* – did I somehow make this happen? And if I did, then how?

Again, that, as they say, is a good question.

Chapter Six
A Telescope into the Past

Charlotte Moberly and Eleanor Jourdain's adventure. Philippe Jullian explains – or does he? Dislocations in time. Jane O'Neil and Fotheringay Church. C. E. M. Joad and the present existence of the past. Lethbridge's ghosts and ghouls. Fields. The ghoul at Ladram Bay. Psychometry. Joseph Rhodes Buchanan. William Denton and The Soul of Things. *Colin Wilson investigates. Life on Mars? Swedenborg. Akashic Records. Rudolf Steiner. Hypnagogia. My dream of Steiner. Discrimination is key. Arnold Toynbee's time slips. Faculty* X. The Philosopher's Stone. *Proust's piece of cake. Knowing in italics. My 'trip to Tibet'. Mental travelling. Present mindedness. Priestley's fish shop. Ouspensky's ashtray. My Indian experience. Making synchronicities happen. Albertus Magnus. Either/or? The purpose of future dreams. The robot. Living dangerously. Time will tell.*

In 1901 Charlotte Moberly and Eleanor Jourdain, two principals of St Hugh's College, Oxford, decided to take a trip to Versailles. While there they got a bit more than what they could expect from their itinerary. While walking in the Trianon park they felt a curious depression and a dream-like sensation for which they couldn't account. What they saw was even more odd. Several people in eighteenth century dress walked past them. They asked two gardeners, similarly attired, for directions. A man who hurried by them advised against taking a certain path. A woman clothed in an old-fashioned dress sat drawing. The two ladies admired some woods, and passed by a rustic bridge that crossed over a ravine with a picturesque waterfall. A man sitting on a garden kiosk greeted them, and they saw a footman in livery emerge from a door in the palace. In the Petit Trianon they watched a wedding party for a time, then left to have tea at their hotel.[1]

It was only after Miss Moberly had written about their visit in a letter that the two compared notes of that afternoon. Miss Jourdain had written her own detailed account, and after reading it the two decided that something odd had indeed occurred. Neither had any particular interest in French history nor in the occult, but the conclusion they came to included both. It seemed that what had happened was that the two of them had somehow been transported back to the Versailles of 1789, just before the French Revolution. Subsequent research suggested that one of the figures they saw, the lady in a light summer dress, drawing, could very well have been Marie Antoinette.

On different occasions the two returned to Versailles. Once, on a visit by herself, Miss Jourdain experienced the odd feeling both had had on their first visit. Two labourers in bright tunics and hoods were loading a cart. Jourdain saw them, but when she looked away for a second and then turned back, they had vanished, although she was surrounded by open space and there was nowhere for them to hide. She also said she heard voices, and the rustling of dresses, but there was no one to be seen.

In 1904, three years after their initial visit, the ladies returned to Versailles and were stunned to find everything different. The woods, the bridge, the ravine, the waterfall, the kiosk were gone. They also discovered that the door to the palace from which they had seen the footman emerge had not been in use for nearly a century; a staircase that led to it had been destroyed some time ago. After this, they were convinced that something very strange had happened that afternoon three years earlier. They began to research the period. It was this study that led them to their conclusion that they had visited Versailles, but the Versailles just before the execution of Louis XVI and his queen. They had experienced a 'time slip'. But instead of jumping ahead into the future, as Dunne did in his dreams, they had slipped into the past.

Just a garden party?

In 1911 Moberly and Jourdain published their account of their experience, calling it *An Adventure*. Soon after the publication, three people who for several years lived in a house overlooking the Trianon told them that they saw the same sort of thing so often that they had become used to it. They ignored it, preferring to 'live in their own century and not in any other.'[2] Some years later, Philippe Jullian, an art historian, seemed to have arrived at a rational explanation for what had happened. In a biography of Count Robert de Montesquiou, Jullian points out that in the early 1890s, Montesquiou, a famous dandy, took a house near Versailles, and used to spend entire days in the park.[3] With a Mme de Greffulhe he often arranged fancy-dress parties for his guests, and this led Jullian to suggest that 'the ghosts' the ladies from Oxford had seen, 'were, quite simply, Mme de Greffulhe, dressed as a shepherdess, rehearsing an entertainment with some friends.' Yet, as Colin Wilson points out, Montesquiou left Versailles in 1894, to move to Neuilly.[4] If what Moberly and Jourdain had seen was a rehearsal for one of Montesquiou's parties, they were seven years too late. Which suggests that if they had seen the rehearsal, that was a 'time slip' in itself. Montesquiou also does not account for the missing woods, bridge, ravine, and other places from the women's account.

Dislocation in time

This is not the only account of people finding themselves suddenly transported into the past. In *The Mask of Time*, Joan Forman has collected many such stories. If, as Dunne tells us, we are always getting a glimpse of the future in our dreams, the opposite seems to be the case too. We are slipping into the past more often than we think. 'The experience of time-dislocation,' Forman writes, 'far from being rare, is very common.'[5]

We may not hear of these experiences for the same reason that we may not hear of experiences of future dreams and precognition: because the person to whom they happen may be afraid of ridicule. Or, equally likely, we may not notice them if they happen to us. I notice my future dreams because I have *learned* to notice them, and I take the time and make the effort to write them down. Someone who doesn't bother to do this may have precognitive dreams every night, but wouldn't know it, not having made the effort to. The Time 2 that Dunne and Priestley tell us is 'above' our everyday tick-tock time, has access to the future and the past. For all we know, in our dreams we may be entering into a past that has nothing to do with us, but which is available nevertheless, on the 'second whorl of the spiral'.

One case of time-dislocation Forman explores involved Fotheringhay Church in Northamptonshire, scene of Richard III's birth and Mary Queen of Scots' execution.[6] Jane O'Neill, a schoolteacher, witnessed a road accident and helped pull injured people from a coach involved in a head-on collision. The experience left her in a state of shock. She was off work for several weeks, and a fellow teacher invited her to recuperate at her cottage in Norfolk. While there Jane began to have odd visions. Images from the accident had stayed with her and had kept her awake at night. Now, at the cottage, these turned into something else. She found herself gazing at pictures and images that appeared for no reason, but sound rather like the kind of images that come in states of hypnagogia. When her friend noticed her distraction, she asked what was wrong. Jane replied, 'I have just seen you in the galleys.' Her friend remarked that this wasn't surprising; her ancestors were Huguenots and were often punished by being sent to the galleys. Jane also saw figures walking in the woods, one of whom she believed was Margaret Roper, the daughter of Sir Thomas More. More odd visions came, but the main event occurred two months later, during a visit to Fotheringhay Church.

In the church, Jane spent some time looking at a picture of the crucifixion on a wall behind the altar. The picture had an arched top, and within the arch was a dove, whose wings followed the curve of the arch. Later, Jane's friend – the one who had invited her to stay at the cottage – read from an essay by Charles Williams (with C.S. Lewis, J.R.R. Tolkien, and Owen Barfield, a member of the Oxford Inklings). In it, Williams said that infinity was often symbolised by a straight line meeting an arch. When her friend mentioned this, Jane said this was what she had seen in the picture in the church: the straight line of the cross meeting the curve of the dove's wings. When she said this, her friend asked 'What picture?' When Jane explained, her friend said that she hadn't seen it. This was odd, because her friend was usually more observant than she was. She was surprised she had missed it.

Jane tried to reach the vicar to confirm what she had seen. The church hadn't had a vicar in years, but the postmistress she spoke with knew the church well and agreed with her friend that there was no such picture. When the two visited the church again a year later, Jane was surprised to discover that it wasn't the church she remembered. From the outside it seemed the same, but inside it was a different story. It was much smaller than she remembered, and the only picture of a dove was of one with its wings outstretched, in a cloud.

Subsequent research revealed that there were other odd stories associated with the church. People had said that they had heard Plantagenet music coming from it; Richard III was the last Plantagenet ruler. Finally Jane wrote to a Northamptonshire historian who informed her that the church she had visited was what was left of a former much larger collegiate church that had been pulled down in 1553.[7] He told her that there was evidence of arched panels on the east wall, behind the altar, where she had seen the picture of the dove. The same evidence confirmed that this included a painting of a dove much like the one she had seen. It appeared that Jane had visited the church as it had been in the fifteenth century.

Lethbridge's ghosts and ghouls

If we accept these accounts as true, how can we account for them? In writing about Moberly and Jourdain's experience, the philosopher C. E. M. Joad – famous at the time as a member of the BBC's 'Brains Trust' – remarked that 'the present existence of the past [though] beset with difficulties of a metaphysical character to which it seems at present impossible to assign any satisfactory solution, [is nevertheless] the most fruitful basis for the investigation of these intriguing experiences.'[8] What did he mean by the 'present existence of the past'?

One person who offered a possible explanation was T. C.

Lethbridge. Along with his work with pendulums, Lethbridge also had a lively interest in ghosts, as he did in quite a few other things. He had seen a ghost during his early years at Cambridge. As he was leaving a friend's rooms, he saw someone entering, and said good evening to him as he left.[9] The next day he asked his friend what his visitor wanted. His friend had no idea what he was talking about. No one had visited him after he had left. Lethbridge then thought that the person he saw was dressed rather oddly, as if he was on a hunting trip. Lethbridge later decided that he had seen a ghost. It was the first of several he would encounter.

On another occasion he encountered a related though different phenomena, what he called a ghoul, traditionally a supernatural creature that feeds on the flesh of the dead. In this case the ghoul occupied a staircase at a Choristers School. He and a friend had met a master who seemed utterly depressed. The reason? 'The ghoul is on the stairs again,' the master said. As a determined empiricist, Lethbridge decided to investigate. As he and his friend mounted the stairs, they felt an odd icy presence, that Lethbridge likened to the sudden shift in temperature he experienced in Greenland when a motorboat he was in moved from brilliant sunlight into the shadow of an iceberg. As they went up, step by step, the 'presence' retreated, until they forced it up two flights of stairs, only for it to return to its original position.

This led Lethbridge to differentiate between a ghost and a ghoul. A ghost is visible and audible, like a picture on a television screen. A ghoul is a 'bad' feeling, a kind of psychic stench, a 'feeling of oppression and horror … often accompanied by the sensation of intense cold.'[10] He was to encounter more than one of each in his career. But how did they get there in the first place?

Pendulums and the past

Lethbridge thought his work with the pendulum offered a possible answer. We've seen that he discovered that the pendulum will respond to abstract ideas. Lethbridge thought that if this was so, then it should also react to emotions. He tested this with some stones he had gathered at Wandlebury camp, an Iron Age fort near Cambridge that Lethbridge had been excavating, uncovering a giant figure cut into the turf. Lethbridge was fascinated with megaliths and other ancient sites, and wrote about them well before they became popular in the 1970s.[11] The stones were most likely used in battle, and when Lethbridge tested them with the pendulum, he discovered that they had a 'rate' of 24 inches (61 centimetres), but that they also reacted very strongly at 40 inches (101 centimetres). He then collected some stones from a beach, picking them up with tongs and placing them in a bucket, so that he wouldn't get his own 'psychic fingerprints' on them. When he tested them the pendulum showed no reaction, neither at 24 inches nor at 40. He then took some of the stones and gave some to his wife and both of them threw them at a wall. When he tested them again, the stones he threw triggered the pendulum at 24 inches; the ones his wife threw got a reaction at 29 inches (74 centimetres). This led Lethbridge to conclude that 24 is the rate for male, while 29 is the rate for female. So, as far as Lethbridge is concerned, differences between male and female seem endemic to reality, and are not a product of culture as many 'progressive' thinkers believe today.

But why had the stones from Wandlebury also produced a reaction at 40 inches? Could it be because they had been used in battle? Lethbridge decided to test this. He lowered the pendulum to 40 inches then thought of something that angered him. Almost as soon as he did, the pendulum responded. So 40 inches was the rate for anger – which, as we've seen, is also the rate for black, cold, death and sleep, all of which seem somehow to form a likely 'cluster', as Kammerer might have said.

But the anger the pendulum had responded to in his test was in his mind. Why did it react to anger that may have been in the mind of an Iron Age warrior, more than 2,500 years earlier? Lethbridge concluded that the anger had somehow been 'imprinted' on the stones. They had 'recorded' it. The emotion was so powerful that it had left its mark on the stones, and it had stayed there for millennia.

Field work

Lethbridge eventually decided that the ghosts and ghouls he encountered were similar 'recordings', imprinted on different 'fields'; this was an idea that Sir Oliver Lodge had proposed many years earlier, in his book *Man and the Universe* published in 1908. I have mentioned only a few of Lethbridge's encounters here as examples; the reader should really go to Lethbridge for the details, and also for the sheer pleasure of reading him. The 'fields' Lethbridge referred to were like those of static electricity or magnetism, and they had a definite shape. He and his wife encountered a ghoul, like the one he had chased up the stairs in Cambridge, on a beach at Ladram Bay. They went there to collect seaweed, which his wife used in their garden. Walking on to the beach, Lethbridge 'passed into a kind of blanket, or fog, of depression.' His wife experienced the same thing. When they mentioned this to friends, they were surprised to hear that they had experienced it too. Ever the empiricist, Lethbridge decided to return and investigate. He found that the ghoul 'had definite limits over which you could step at a single stride.'[12] It occupied a limited area and was especially powerful in warm, damp weather.

The source of the ghoul-field, Lethbridge believed, was a 'terrible mental strain'. It was caused, he believed, by someone struggling with the thought of suicide. His wife had felt it when standing on a cliff within the field. The thought of jumping onto the rocks below came to her as she stood there; it was as if someone were suggesting it to her. Lethbridge thought that someone had either jumped off the cliff, or wanted to do so, and that the struggle going on within them had been 'recorded' by the field he had entered. And taking a leaf from Priestley's FIP, Lethbridge even suggested that 'since time does not seem to follow its normal course in these things', the thoughts he was picking up might even have been those of someone who in the future will struggle with the idea of killing themselves on that spot. He adds that a few years after writing about the ghoul at Ladram Bay, someone had thrown themselves off the cliff. Was this the future suicide whose thoughts had somehow imprinted themselves on the area *before* the act? Or was the suicide a depressed individual who followed the suggestion made to Lethbridge's wife about tossing herself on to the rocks?

In either case, the idea is that human emotion can impress itself on certain 'fields' and that the recording of the emotion can be picked up and played back in the mind of people entering it. Lethbridge suggested that there are different fields associated with different landscapes and terrains, and that the nymphs, dryads, naiads, oreads, nereids, and elementals of ancient mythology were personifications of these fields.

How far does this go in explaining Moberly and Jourdain's 'time slip'? Lethbridge sees ghosts as recordings, rather like what we see on television. In this case, rather than infect a field with a kind of psychic poison, the image of an event is somehow recorded on the field in the way that the same image can be recorded on film. Our minds play back the recording but, just as with television, we don't 'interact' with them.

Jourdain and Moberly did interact with their 'ghosts'; they asked directions and they received advice, so it seems that something more than a recording was involved. But for our purposes this is irrelevant. The important point is that a past event can leave its 'mark' on the world around it, and if we are in the right state of mind, we can 'play it back'.

Psychometry

A similar idea is at the heart of psychometry, mentioned earlier. This is the strange ability to learn about the history of an object, merely by holding it. We can say that when Lethbridge used his pendulum to find the rates of different objects, he was performing a kind of psychometric reading of them.

The term 'psychometry' – psyche (soul) metre (measure) – was coined in the early nineteenth century by Dr Joseph Rhodes Buchanan, a physician, mesmerist and student of phrenology, the psychology of bumps on the skull. Buchanan learned of someone who said he could detect brass in the dark, simply by touching it. It produced a metallic taste in his mouth. Buchanan was intrigued and he began to test his students, wrapping different substances in paper so that they wouldn't know what they were, and then handing them to them. He found that not only could they detect brass, they could do the same with other substances: salt, pepper, sugar. And not only substances. When one of Buchanan's students touched the stomach of someone who was ill, he became ill too. And just as Lethbridge would discover with the pendulum a century or so later, psychometry worked with things more abstract. Buchanan gave one of his 'sensitives' some letters from his files. The sensitive was able to tell him a great deal about the writer, simply by holding a letter. It seemed the writer's *personality* was in the letter in more ways than one.

Buchanan carried out hundreds of tests. He concluded that an individual 'leaves the impression … of his mental being upon the scenes of his life,' and that by learning how to decode these, we will have found a 'new clue to the history of our race.' 'Unrecorded ages' were now available to us, because objects from the past were 'still infused with the spirit that produced them.' Through psychometry the 'living realities' embedded in these ancient objects could be revealed. Buchanan summed up his insight in an enthusiastic pre-echo of C. E. M. Joad's conclusion about the adventure at Versailles: '*The Past is entombed in the Present!*'

The soul of things

Buchanan's follower, William Denton, a geologist at Boston University, was even more enthusiastic about this new science, as his still readable book *The Soul of Things* shows. He conscripted his sister into his experiments. She turned out to be one of Buchanan's 'sensitives'. He wrapped a piece of limestone taken from the Missouri River in paper, didn't tell her what it was, and handed it to her. She told him she had an impression of water, a river, shells – the limestone contained tiny shells – and described the spot where Denton had found the stone. A piece of volcanic lava, also wrapped in paper, produced a vision of an 'ocean of fire pouring over a precipice.' The lava had come from Kilauea, an active volcano in Hawaii. To eliminate telepathy, Denton jumbled his uniformly wrapped objects so that even he didn't know which one he was handing his sister. Holding a meteorite, she had a vision of outer space. With a piece of bone taken from limestone, she saw a prehistoric world.

With the pendulum, Lethbridge was able to detect the anger an Iron Age warrior had impressed upon a stone. But as psychometry seemed to show, with that stone a sensitive could somehow *see* the past that had been impressed on it.

But Moberly and Jourdain did not say that they saw Versailles in 1789 when they touched anything in particular during their visit. And the people who lived near the park and who had got used to seeing 'another century' out of their windows, were not in contact with anything either. Their time slip was related, but different. But again, we're not looking for an explanation; we're simply looking at different accounts of the 'present existence of the past'.

A telescope into the past?

In *The Psychic Detectives*, his book about psychometry, Colin Wilson points out one problem with this otherwise miraculous 'telescope into the past'.[13] It is unreliable, or at least not always reliable. As Denton continued his experiments, his subjects were able to differentiate among different periods in an object's past, to 'tune in' to different eras as we would a radio broadcast. And their descriptions of what they saw became more and more detailed. One sensitive produced a remarkably accurate description of ancient Pompei. Another gave an account of the home of the Roman dictator Sulla. From a piece of hornstone from the Mount of Olives, one of Denton's subjects described ancient Jerusalem.[14]

The results of psychometry, and discoveries like Lethbridge's, lead Wilson to consider that we live in an 'information universe'.[15] It may be that we live in an 'information age' more than we think.

Yet this seeming accuracy was accompanied by some visions that could not be corroborated, or were later shown to be simply wrong. Denton's son Sherman became a sensitive, one of the most powerful he had tested. But in the third volume of *The Soul of Things* Sherman's psychometric skill turned itself outward, to the stars, at least to our neighbouring planets. He gave accounts of Mars, Venus, and Jupiter that have subsequently been shown to be wildly off the mark, if they were believed at all in 1874 when they were published. Venus had trees like giant toadstools, and was peopled with weird half-fish, half-muskrat creatures. The inhabitants of Mars were blonde and blue-eyed, had four fingers and wide mouths. Those on Jupiter were also blonde, and could fly. Today we know that Mars is cold and barren, Venus is too hot to support life, and Jupiter is mostly a gigantic ball of gas. What happened?

This odd mixture of accuracy and what strikes us as fantasy is, unfortunately, not uncommon among visionaries. Swedenborg was taken by angels to visit heaven and hell, but he also made trips to Mars, Jupiter, the moon, and other planets. If Swedenborg's accounts of heaven are not enough to put off some modern readers, what he tells us about the inhabitants of these planets surely breaks the 'believability barrier'. Martians have faces that are half black and half white and they communicate via telepathy. The people of the Moon have booming voices and they speak through a kind of burping. Jupiterians, who are rather like us, are kindly, noble, highly moral, virtuous and naked most of the time.

We may chuckle at these descriptions, but Swedenborg's accuracy in other matters was well attested. While in Gothenburg, he knew that a fire had broken out in Stockholm, three hundred miles away. He described it accurately and even knew when it had stopped, just a few houses from his own. There is also the story of Swedenborg informing the queen of Sweden of some information that only her dead brother could have known; Swedenborg acquired it by speaking with her brother on one of his visits to heaven. On another occasion, he was able to save a widow from having to pay for an expensive silver tea service that she knew her husband had already paid for. She was unable to locate the receipt until Swedenborg told her it was in a secret drawer in her husband's desk. Her husband had told Swedenborg about it on one of his visits to the spirit world. The widow found the receipt where Swedenborg said it would be.[16]

Akashic records

Another area of visionary ambiguity involves what is known as the Akashic Record, a kind of chronicle of human and cosmic evolution, imprinted on what occultists call the 'astral light'. This is a kind of immaterial ether on which, as Madame Blavatsky put it, 'is stamped the impression of every thought we think.' From Lethbridge's perspective, the Akashic Record is a kind of field enveloping the entire cosmos. Blavatsky claimed to be able to read this, as did other theosophists, like Rudolf Steiner. In my book on Steiner I give some accounts of how Steiner did this. I also talk about how, like Jung and Swedenborg, Steiner was a well-practised hypnagogist. While Lethbridge and the psychometrists spoke of past times that are part of the historical record – except, of course, for the flights to Mars and Venus – what Steiner, Blavatsky, and other readers of the Akashic Record, such as Edgar Cayce, speak of are remote prehistoric periods, ages in humanity's dim past during the days of Atlantis and Lemuria – if, of course, they existed. If it was impossible to verify Sherman Denton's accounts of Mars in 1874, we know they are inaccurate now. So far, we have no way of verifying anything Steiner or Blavatsky tells us of these lost continents from ages ago.

Madame Blavatsky may have been an explosive madcap who didn't give a hoot about verification and would brush aside any such concerns with a wave of her cigarette. But Steiner was sobriety incarnate – rather as Swedenborg was – and the success of his ideas in practical matters such as agriculture, medicine, architecture, education, the arts, and other fields, suggest a mind firmly planted in the real world. I began to give him the benefit of the doubt after reading some of his criticisms of Kant. I was studying philosophy at the time and Steiner's remarks about epistemology struck me as insightful and well-founded. After that I was willing to keep an open mind about some of the other things he said.

As I point out in my book, Steiner seemed able to shift his consciousness into the sort of reverie associated with hypnagogia and this became for him, as it did for Jung, a form of 'active imagination'. (We remember Koestler's remark about regressing from 'articulate verbal thinking to vague, visual imagery.' Steiner himself talks about a 'picture thinking' that preceded our language based thought.) Steiner was able to *see* Atlantis and Lemuria in the way that Denton's sensitives 'saw' ancient Rome.

But as we've seen in our look at 'future dreams', the accurate information about what is to come is usually mixed up with images from the past, current worries, and the creative productions of the 'dream artist' who, as I mention at the beginning of this book, can create scenarios that are often *more real* than our everyday experience. I'll mention here that in one dream, which I don't have space to recount in detail, I believe that I was visited by Steiner, one of the 'meetings with remarkable men' I mentioned earlier. Was it *really* Steiner? I don't know. But I do know that if it wasn't, my dream artist was certainly able to create a dream Steiner that struck me forcefully as the genuine article.

We know that dreams speak in symbols, metaphors, images, analogies, not in straightforward factual prose. We also know that some fairly incredible things happen in dreams: we fly, walk through walls, speak with the deceased or with animals. And just as some accurate information about the future or something else can be communicated in a dream, it is also the case that it is often difficult to sift through the flights of creative fantasy to reach the nuggets of fact. In *A Secret History of Consciousness* I point out that while it may be impossible to confirm any of the *details* Steiner provides about human consciousness before the historical record, his ideas about the *nature* of that consciousness strike me as worthy of serious attention. It is easy to accept everything dreams offer as true, as the credulous 'new ager' does, or to reject them entirely as psychic rubbish, as hardnosed rationalists do. What is difficult is to patiently discriminate between what might be valuable and what seems utterly fantastic. The same is true, I think, of something like Steiner's or Swedenborg's visions.

Arnold Toynbee's time slip

Yet there is another kind of 'telescope into the past' that is less open to immediate rejection by sceptics. A 'time slip' that Wilson writes about in *The Occult* and other books is one that happened to the historian Arnold Toynbee in the early twentieth century; it was one of several that Toynbee experienced. As he recounts in the tenth volume of his *Study of History*, in May of 1912, Toynbee visited the ruins of the citadel of Mystra in Greece, near ancient Sparta, which had been left desolate since the invasion of highlanders in 1821 had decimated its inhabitants. As Toynbee sat and meditated on the 'cruel riddle of mankind's crimes and follies', he had a sudden and unmistakable sense of the *reality* of what had happened on that April morning at the start of the Greek war of independence. Toynbee of course knew that the attack had taken place. But now, sitting, gazing at the ruins with the slopes of Mount Taygetus in the distance, Toynbee *really* knew the massacre had taken place. It was as if he had been sent back to it, as Moberly and Jourdain seem to have been sent back to Versailles on the eve of the revolution. But there was a difference. Toynbee was an historian, unlike the ladies from Oxford. He knew a great deal about the battle, all the facts. But now the facts had 'come alive'.

Toynbee had similar experiences. In one he was 'transported, in a flash, across the gulf of Time and Space from Oxford in AD 1911 to Teanum in 80 BC' where he witnessed one of the outlawed leaders of the Italian Confederacy commit suicide when his wife refused to help him. Another time he was sent back to accompany the Spaniards as they first reached Tenochtitlan; on another he was with the first crusaders to reach Constantinople – which, sadly, they sacked. All of these brief 'time slips' – they lasted only a few moments – culminated with a vision of *all* history, as he walked along Buckingham Palace Road near Victoria Station. On this occasion he found himself 'in communion not only with this or that episode in history, but ... with all that had been, and was, and was to come', which sounds rather like a glimpse of Steiner's Akashic Record.[17]

Toynbee was no occultist. He entered no trance or altered state when he had these experiences. What had happened was that the *knowledge* of history he possessed was suddenly illuminated by his imaginative grasp of the *reality* of the facts making up that knowledge. In speaking of Mystra, Toynbee mentions the 'effect upon his imagination of a sudden arresting view of the scene.'

We may feel something like this when we first visit somewhere we've always wanted to see, a sense of 'What? Me, *here?*', especially if it is somewhere we have some knowledge of. But when his vision of all history came to him, Toynbee was surprised that such an 'incongruously prosaic scene' as Victoria Station should have been the setting for his 'mental illumination'.[18] But if history is real, then all of it is, and it is everywhere around us.

Faculty X

What had happened, Wilson says, is that Toynbee had a sudden moment of what he calls Faculty X, our unrecognised ability to grasp 'the reality of other times and places', that I mention at the beginning of this book. Not the mere *knowledge* of these other times and other places – Toynbee had that in spades – but their *reality*. This is another example of, as C. E. M. Joad had said, 'the present existence of the past.'

We can say that what had happened in Toynbee's time slips is that the left brain knowledge he had of history was suddenly made 3D by his right brain supplying a good dose of reality – the *meaning* that the abstemious left brain usually edits out. His imagination, we can say, made what he knew real. In other words, he 'realised' it.

Wait a minute. The imagination making something real? Isn't that wrong? Isn't imagination the *opposite* of reality? If by imagination we mean fantasy or 'make believe', then yes. But that is not what imagination is really about. Imagination is the ability to grasp realities that are not immediately present – which is another way of defining Faculty X, the ability to grasp the reality of other times and places. Either way, imagination is about reality. Priestley would agree. He believed that imagination, 'that mysterious, multi-dimensional faculty', which is 'not yet pinned down and labelled', is the 'first dimension' of the new reality our moments out of time reveal to us.[19]

The Philosopher's Stone

In his novel *The Philosopher's Stone*, Wilson writes of scientists who, by operating on the brain's pre-frontal lobes, are able to induce a permanent opening of consciousness, so that the reality which is usually edited out is made available. (He was unaware of split-brain psychology when he wrote it.) At one point the hero finds himself resting on the lawn of a Tudor house. Gazing at a shallow ditch, he suddenly felt with 'complete certainty' that the ditch was the remains of a moat. He then turned to the house and using 'intelligent speculation' he tried to imagine what it would have been like in the 1500s. He speaks of imagination as a kind of 'inward seeing', that enabled him to 'grasp the reality of factors not actually present to the senses.'[20] Wilson's hero is then treated to an extended time slip, like the one Moberly and Jourdain experienced. But in his case it didn't just 'happen'. He made it happen. How? By using his imagination.

The kind of imagination Wilson's hero used is 'a power to reach out beyond present reality just as radar can penetrate clouds.'[21] What a radar beam detects is 'really' there, but it is not immediately available to our senses. Our inner radar works in exactly the same way. What Toynbee did involuntarily at Mystra, Wilson's hero does intentionally. This is a crucial difference; 'intentionality', based on the philosopher Edmund Husserl's central insight that 'perception is intentional', not passive, is at the essence of Wilson's work.[22] It is a reaching out and *grasping* of reality, rather than a passive, mirror-like reflection of it. The more we 'intend' – that is, the more attention and will we put into our perceptions – the more reality we will see. By putting more intention into our attention, we can open the doors of perception and allow more reality into consciousness, without getting overwhelmed by it as can happen through using 'mind-altering' substances.

Throughout the book, Wilson's scientists have several 'time slips'. Yet, unlike Joan Forman, who believes that 'There seems to be no known method of controlling the genuine "slip" into the past or future', Wilson believes there is.[23] But for him the machine that can enable us to do this isn't a product of technology, nor does it depend on the vagaries of elementary particles. Time travel, Wilson tells us, is a 'purely mental faculty.' H. G. Wells' Time Machine is 'the human mind itself.'[24]

The madeleine moment

Another example of Faculty X occurs at the beginning of Proust's *Remembrance of Things Past*, when Marcel tastes the madeleine and is suddenly reminded, not of the *fact* that his aunt used to give him a taste of her madeleine dipped in herbal tea when he was a boy, but of its *reality*. I can't emphasise this enough. Here language works against us. In order to make my point, I have to say that although Marcel *knew* that he had holidays in Combray with his aunt, the madeleine has made him *really* know it. He now knew it in italics, we might say. Surely, as George Steiner pointed out, our dictionaries lag behind our needs. I should point out that Wilson wasn't the only one to recognise the differences between these kinds of knowing or the kind of memory involved in them. In *Janus: A Summing Up*, Arthur Koestler contrasts 'abstractive memory', our usual kind, which provides 'the dehydrated sediments of experiences whose flavour has gone', with 'spotlight memory', which recalls scenes and episodes with 'almost hallucinatory vividness.'[25] Koestler and Wilson are speaking of the same thing.

We experience something like a time slip whenever we become deeply absorbed in a book. Under the right conditions, and with the right book, we can become what William Blake called 'mental travellers'. When researching my book on Madame Blavatsky, I had to read quite a bit about Tibet. I can remember finishing my work for the day, and then heading out of my flat for a walk. For the first few moments I was surprised to be here, in North London, and not somewhere in the vicinity of the Tashil Lhunpo monastery, where I had been all day. My neighbourhood was real, no doubt. But I had 'mentally travelled' to another, equally real location in space and time, and the dissonance between the Himalayas and West Hampstead took some time to dissipate. (I used to feel the same thing when I was a boy, coming out of a Saturday matinee at the cinema and having reality thrust in my face.) How had I got on the other side of the world? Through the 'inward seeing' of my imagination. When the imagination is able to 'transport' us in this way, the difference between this form of 'inward seeing' and the paranormal sort we call clairvoyance or remote viewing seems one of degree, not kind. And if we can view remotely in this sense, why not through time as well?

J. B. Priestley's fish shop

What seems necessary is some interest, some knowledge, and that empathy and absorption that characterise getting 'deep into a book'. We may not know we are doing it, but when we become so absorbed we are involuntarily increasing our intentionality. That is why it often strikes us that what we are reading about seems more real than the reality around us. People to whom this happens are often called 'absent-minded'. They are nothing of the sort. Their mind is 'present', but not in the immediate reality around them. If they are 'absent' from it, they are certainly 'present' elsewhere, most likely in a 'reality' more interesting than the bland immediacy before them.

So while I may not be able to transport myself at will to Versailles before the revolution or to a church in the fifteenth century, I can travel in time in another way. As Wilson says, I can experience the reality of other times and places mentally, in the same way that I can get on a flight to a foreign country.

J. B. Priestley had an experience of Faculty X, although he didn't call it that. It didn't transport him to any particular time and place, but it did widen his grasp of reality and take him 'outside' of time. Priestley writes of finding himself outside of a 'large, fine fish shop.' As he looked at the 'scales and fins and the large round eyes, looking indignant even in death', he felt that he lost himself and any sense of 'passing time' in a 'vision of fishiness itself, of all the shores and seas of the world, of the mysterious depths and wonder of oceanic life.' He called this experience, and others like it, the 'aesthetic feeling', because he believed that 'real poets ... must always be enjoying such selfless and timeless visions.'[26] He says that it can come to him merely by hearing the words 'France' or 'Italy' or 'eighteenth century'. What they bring is 'a feeling for the immense variety, richness and wonder of life on this earth.'

Ouspensky's ashtray

These 'aesthetic feelings' are not full-on mystical experiences, but some mystical experiences seem an intensified version of what Priestley is speaking about. While experimenting with nitrous oxide, Ouspensky found himself looking at an ashtray and suddenly being overwhelmed by everything to do with it: the history of smoking, tobacco, mining – the ashtray was made of copper – fire; 'an infinite number of facts' having to do with the ashtray assaulted Ouspensky and he realised that 'we do not understand the simplest things around us.' Ouspensky tried to capture some sense of his insights before they faded. The next day he saw what he had written: 'A man can go mad from one ashtray.'[27]

William James had a similar experience, induced by nothing more than a conversation. During it, he found that he was suddenly reminded of something, and this reminded him of something else, and then of something else, at such a rapid rate that he found himself 'amazed at the sudden vision of increasing ranges of distant facts of which I could give no articulate account.'[28]

We can say that in Priestley's, Ouspensky's, and James' case, some of the reality that our hyper-efficient left brain edits out was allowed into consciousness, with the result that they each briefly stepped out of 'passing time' and had entered something like Dunne's Time 2 – at least a dimension of time in which what would normally require lengthy study (left brain) came to them all at once (right brain). Priestley's 'aesthetic feeling' was not overwhelming, just as Proust's madeleine moment wasn't. But with Ouspensky and James, the fact that they could give no 'articulate account' of what happened suggests that the verbal left brain was left stunned by the amount of information coming to them. If we live in an 'information universe', then even as humble an item as an ashtray can contain within it more information than we might assume.

I had a moment of Priestley's 'aesthetic feeling' – or Wilson's Faculty X – some years ago, when I was living in New York, around the same time that I was first reading Dunne. I had been out for a walk with friends and we had returned to my apartment. At that time I had a globe, and it was placed so that it was one of the first things I saw when I entered the flat. I remember the sensation of walking into the warm room after being on the cold streets, taking off my coat and scarf; I can even see my glasses fog over for a moment. Then I saw the globe, the eastern hemisphere; India, in fact. I remember saying to myself – or hearing – the words 'the sub-continent'. And then it was as if I was in a theatre and was watching a film, but a theatre suspended miles above the 'sub-continent', and a film that covered its entire history in a few seconds. I saw armies, massive migrations, tremendous battles, dense jungles, temples, mountain ranges, everything to do with the phrase 'the sub-continent'. And at the same time I was in orbit above India, as if my position in relation to the globe in my apartment had been matched to the real planet in outer space. It was a sensation of hovering not only over the actual physical 'sub-continent', but of seeing its history trail away into the distance, of seeing a great procession stretching backwards in time. It lasted all of a few seconds, but I have no doubt that in that time I had gained an overview of the objective reality of India that I might begin to approach only after years of study. I have had other moments of Faculty X, but nothing else quite like this.

Making synchronicities happen

What Faculty X, Priestley's 'aesthetic feeling', and other similar experiences tell us is that we somehow have more power over our consciousness than we realise, and that what we take for 'normal' reality is anything but. It is a severely edited version tapered to our practical needs and purposes, with nearly everything of non-utilitarian value leached out. And once we grasp this intellectually – once we know this in the same way that we know the earth goes round the sun – we can then start to learn how we can open the aperture of consciousness to allow more reality in, just as we open a curtain to let in more sunlight.

One of the effects of the intention to understand the inner mechanisms of consciousness is a change in consciousness itself, or at least a change in what it is conscious of. Koestler was subject to a 'meteor shower' of coincidences while researching his book on coincidence. Wilson and others can tell similar stories. I can myself and so, probably, can you. It seems that talking about coincidences, or synchronicities, attracts them, so if you start collecting them, you will soon have quite a few on your hands. Wilson found that when he was in a purposive state of mind – say when writing a book – good coincidences and synchronicities would happen; the library angel would be on duty, and material and references he needed seemed to be on tap. And the opposite was true. If he was distracted, upset, gloomy, things went wrong and he became accident prone. By all accounts Wolfgang Pauli was prone to emotional upheavals and led a wild, even dissipated life; this was why he came to Jung for help. His inner conflicts may have been the cause of the lab disasters he left in his wake. I'm convinced that in my own life, one or two light bulbs have burst because of a bad mood; possibly a window or two cracked as well. If reality is fundamentally 'psychoid' as Jung says, and at some level our inner and outer worlds interpenetrate, then Jung may have been more right than he would have liked to have been, when he spoke of a 'pre-existent psyche which organises matter.' If perception is intentional and consciousness reaches out to *grasp* reality, who knows how far that grasp might reach?

Jung believed that synchronicities involved what he called an *abaissement du niveau mental*, a lowering of the mental threshold, a loosening of concentration and attention, as happens as we fall asleep. We know that precognition and other paranormal phenomena are partial to hypnagogia. But Wilson discovered the opposite also to be true: that a highly energised, intentional state seems to attract them. Jung himself is an example of this, when he manifested the poltergeist in Freud's bookcase during their argument. But Jung didn't want to accept that he, or anyone else, could make a synchronicity happen. That would have been magic.

It was to a magician that Wilson referred when questioning

Jung's insistence on the *abaissement du niveau mental*. The thirteenth century alchemist and sage Albertus Magnus – whom Jung certainly knew of – spoke of a 'certain power to alter things' residing in the 'human soul', which enables it to 'subordinate other things' to itself. He speaks of a 'great excess' of emotion being able to 'bind things together' and 'alter them in the way it wants.'[29] 'Binding things together' sounds rather like one of Kammerer's 'clusters' and 'altering them' the way the soul wants sounds rather like synchronicity. Albertus Magnus declares that 'Whoever would learn the secret of doing and undoing … must know that everyone can influence everything magically if he falls into a great excess.'

Jung also tells us that synchronicities occur when an archetype is 'activated', which seems rather like the 'great excess' Albertus Magnus speaks of. This seems to contradict the 'lowering' of consciousness that Jung insists on. But what Jung means, I think, is a lowering of the ego, or left-brain critical consciousness, that overrides the intuitions and communications from the right brain, which, we've seen, is associated with synchronicities. A great excess of emotion, as Jung felt towards Freud during their argument, has the same effect. If Freud's dismissive attitude hadn't infuriated Jung, the poltergeist most likely would not have turned up.

And if the soul can alter things in the here and now, what's to stop it from altering them in the future, or even, as Ouspensky believed, in the past?

Either/or?

This seems to suggest a possible way out of the 'precognition/ free will paradox' that future dreams present us with and which I mentioned at the start of this book. Like the paradoxes of Zeno, which have an arrow never reaching its target and a tortoise outrunning Achilles, it may be more of an apparent paradox than a true one. These paradoxes, we've seen, are really a product of the left brain's 'either/or' attitude towards reality, which insists on the 'excluded middle', that is, something being either A or not A, yes or no, black or white. But this insistence on one or the other is dissolved with synchronicities, which allow for an 'included middle', a liminal realm that is the ontological equivalent of the hypnagogic, a 'both/and' with which the right brain is at home, Jung's 'transcendent function'.

Priestley had no patience for the 'either/or'. 'If we are to go forward into a higher, less abstract, richer more fundamental reality,' he writes, 'the *either-or* simply will not work.' As he explains:

Either men are mortal or immortal, eh? No. Either you
must accept determinism or freewill. No, you mustn't.
Either the future exists or it doesn't exist. No, it both exists
and doesn't exist. Either personality is real or an illusion.
No, it is both or, if you will, neither.[30]

We've seen that Priestly considers the future to be multiple, and that the knowledge of it that comes to us in dreams may be knowledge of one, perhaps the most likely future. But it is not the *only* one. The idea that there are multiple possibilities in every moment is something both science and esoteric philosophy agree on; in different languages, John Wheeler and Ouspensky are saying the same thing. And, as Priestley says, because the future can be seen, it can be changed. Why then are there fewer examples of a precognitive dream enabling someone to change the future than of future dreams 'coming true'?

One answer is that the majority of future dreams are about trivial things. In these cases, as Stan Gooch pointed out, *that* they were precognitive is more significant than *what* they were precognitive about. A mother saving her child from drowning is not trivial, but most future dreams are not concerned with such matters; at least mine haven't been. We might also ask if the future that comes in dreams is perhaps one that should be actualised, that it is a desirable future, aside, of course, from catastrophes. I would also suggest that the purpose of future dreams is simply to keep us on our toes, to shake us out of our usual drifting approach to life, to make us *wonder*, rather as synchronicities do. And this leads to another reason why we tend to accept a future that is unalterable, rather than a potential but still indeterminate one.

The robot

If we were already sensitive to wonder, if our 'indifference threshold' as Colin Wilson calls it, were not so high, and if we were less under the thrall of what he calls 'the robot', we would perhaps not need to be stunned into an awareness of how strange and miraculous reality really is. As it is we do need this, because for most of our existence, 'we' are not really living. Our robot is living for us. It is an evolutionary labour saving device that has enabled us to master the world – for good or ill is debatable, yes, but either way, we have mastered it. Animals have robots, but none as versatile as ours. When we learn anything, how to type or speak French or ride a bicycle, it is a laborious process *until* our robot steps in and starts to do it for us. The robot is excellent at this, and because it is we are able to move on to a higher level of purpose. We can then think of *what* we want to type or say in French or *where* we want to cycle to, without thinking about *how* to type, speak French, or ride a bike. The robot does that for us. Without this automatic pilot, we would never get beyond the first stages of learning how to do these things. In effect we would never learn them at all. Learning *is* precisely handing these operations over to the robot.

But as with the left brain, our robot does its job *too well*, and it does things that we would rather do ourselves. It could be listening to Mozart, appreciating scenery, or enjoying a meal. 'We' want to do these things, but like an ever-attendant valet, our robot is there, doing them for us. When it does we wonder why the *Jupiter Symphony* bores us or why that sunset fails to move us. Just as the left brain edits out whole dimensions of reality that have no practical value for it, our robot's mandate is to perform repetitive functions with as little energy as possible. Both it and the left brain are firm conservatives, tightwads on our energy supply. And as we allow more and more of our life to be lived for us by the robot, the world becomes less and less 'interesting'. It loses 'newness', and we become subject to the discouraging levelling of 'passing time', with 'one damned thing following another' in numbing regularity, that 'being-towards-death' that the existentialists speak of. Tick-tock. If consciousness is intentional, robotic consciousness is as unintentional as it can get. In 'The Rock', T. S. Eliot asks, 'Where is the life we have lost in living?' It is in the hands of the robot.

But the robot isn't a villain. We need it just as we need the left brain. Both are absolutely indispensable. We would have never left the trees without them. Aldous Huxley recognised this when he said that if everyone took mescaline, there would be no wars, but there would no civilisation either. Mescaline and other drugs give the robot the day off; but when the robot is on holiday the dishes pile up. Bohemian daydreams aside, that is not a real option. No, we need the robot. And there are times when we and the robot work together. When a musician is inspired, the left brain gets an intuition from the right brain; the robot obliges, and a passage they have played a hundred times is suddenly bold and new. This is true for all of us: when the three work together, even the simplest things can be a source of delight. The problem is not so much that the robot has usurped power as that we have allowed it to infer that we would like it to do everything for us. And the robot is happy to oblige.

Live dangerously

Wilson discovered that crisis and inconvenience can free us from the robot; this is what drove his 'outsiders' to 'live dangerously', as Nietzsche suggested and as I told my audience at the Theosophical Society. Crisis forces us to make an effort; the robot gets the message and hands the steering wheel over to us. Yet once the crisis has passed, we hand it back to the robot. We may not want to do this, but out of habit we do and we are surprised when things seem dull again. Given that we have created civilisation precisely in order to minimise crisis, there seems something absurd in going out of our way to look for it, rather like putting a gun to our head in order to remember how much we want to live, which is something some 'outsiders' did. But there is also another way. It is simply to make efforts to be more intentional. To do things with purpose and to consciously put more into doing them than is necessary. In this context Wilson often reminds of us Hermann Hesse's remark in *The Journey to the East*, that 'a long time devoted to small details exalts us and increases our strength.'

Devoting a long time to small details strikes us as a bore. Our robot hears this and says, 'Here, let me.' If, instead of allowing it to take over, we put more intention into doing it, whatever it is, and persist in this, an odd thing happens. We can find that we *enjoy* doing it, as I discovered when recovering from an operation forced me to put more attention into each step I took. We have convinced the robot that, like a crisis, *this* is important, and so he lets us do it. The 'enjoyment' comes from this. We are living for a change, not having our life lived for us. Walking on crutches is not particularly enjoyable. But the *attention* I put into doing it made it so. What was drudgery now seemed delightful. Now that I know this, I can make the same effort, without the crutches.

Yet much of our life is repetitive and so the robot has plenty to do. We take the same train to work each day, or drive on the same highway, or take the same bus. The railway, highway, and streets are the same, and what we see varies little. So we can say that much of our life is more or less predetermined. Which is not wholly a bad thing; an absolutely undetermined life would be chaos and soon prove unliveable. But luckily we do not have to choose between one or the other. *We* can be different, even if things aren't. We can decide to think about time and its mysteries as we head to work, or we can sink into the same repetitive thoughts as we had the day before. We can decide to record our dreams, or ignore them as irrelevant. We can decide what to do with our inner world. We can make an effort to grasp something of its reality, or we can let our robot do our thinking for us. One thing I can say is that if we do make this effort, what strikes us as boringly familiar can suddenly seem oddly strange, and what we think we know already can reveal a new face and convince us that, as Ouspensky discovered about his ashtray, we didn't really know it at all. Synchronicities, precognitive dreams and other taps on the shoulder by our guardian angel, nudge us towards this awareness, but we can help them along.

Wilson believed that 'someone who had achieved a perfect level of collaboration between the right and left hemispheres ... would be able to slow time down or speed it up at will.' He also believed that if that same someone 'could reach some entirely new level of delight and concentration, time would virtually disappear.'[31] As we've seen in this book, such collaboration between our brains and such concentration has been achieved at different times and with different people, so it is entirely possible. We can only imagine what human life would be like were it to become a more permanent feature of consciousness.

Time, as they say, will tell. But that is one future dream I wholeheartedly wish to come true.

London, April – June 2020.

Acknowledgments

Many people helped make this book possible. I'd like to thank Suzette Field and Stephen Coates for inviting me to speak as part of the 'Borderlands of Sleep: London Dreaming' series of events and for letting me indulge myself and entertain my audience with accounts of some of my dreams. I'd like to thank Dean Radin for inviting me to speak at the Real Magic seminar at the Omega Studios in Rhinebeck, NY, and Alex and Alyson Grey and Regina Meredith, with whom I shared the podium, for a fascinating weekend. I'd like to thank Helene Artz for making possible my stay in Montreal, much of which was devoted to reading Eric Wargo's *Time Loops*. Ray Grasse's stellar insights were of much help, as were Malcolm Rushton's regarding some dreams that I did not get to in this book. James Hamilton was again a patient listener. I would also like to thank Ally, the Queen of Synchronicity, for looking at a part of the work while in progress. Once again, I am indebted to the staff of the Swiss Cottage and British Libraries, without whose help my research would not have been completed before coronamania hit town. My appreciation goes out to the frontline workers who kept life going after it did. I would also like to thank my ex-wife and other friends and partners, without whom some of the experiences I discuss in this book would not have been possible. I trust they will not mind that I have shared them with my gentle readers. And lastly, I would like to thank that thoughtful member of my audience at Brompton Cemetery, without whose tweet about hedgehogs I might not be writing these words.

Praise for Gary Lachman

'One of the leading students of the western esoteric tradition, Lachman has published critical studies of Swedenborg, Madame Blavatsky, Aleister Crowley, Rudolf Steiner, P.D. Ouspensky and Jung – and he has done so without being raptly worshipful or casually dismissive.'
The Washington Post

'A cracking author.'
Magonia Review of Books

'Lachman is an easy-to-read author yet has a near encyclopaedic knowledge of esotericism and is hence able to offer many different perspectives on the subject at hand.'
Living Traditions magazine

Praise for *Lost Knowledge of the Imagination*

'Very important.'
Philip Pullman

'Lachman manages to make basic concepts in esoteric philosophy and history lively as well as readable.'
Kirkus Reviews

'Lachman creates a history of ideas that fascinates and excites.'
New York Journal of Books

'An excellent book – scholarly but eminently readable by anyone seeking appreciation of the spiritual.'
Alister Hardy Society

Praise for *The Caretakers of the Cosmos*

'This is one of the most stimulating and significant books on the subject in years. *The Caretakers of the Cosmos* is an essential work for all who are curious about what makes us uniquely human, and about how we can all participate, each in our own way, in the creation of a fuller and more satisfying world.'
Parabola magazine

'Lachman's depth of reading and research are admirable, and he weaves the story well, developing what is becoming increasingly obvious – that all of human history reflects the evolution of consciousness.'
Scientific and Medical Network Review

Endnotes

Introduction: Of Hypnagogia and Hedgehogs

1. http://londonmonthofthedead.com/hypnagogia.html.
2. See especially *A Secret History of Consciousness* (Great Barrington, MA: Lindisfarne Books, 2003), pp. 85–94.
3. F. W. H. Myers, one of the founders of the Society for Psychical Research, proposed 'hypnopompic' for the liminal state as we awaken, complementing L.F. Alfred Maury's earlier coinage of hypnogogic.
4. Andreas Mavromatis, *Hypnagogia: The Unique State of Consciousness Between Sleeping and Waking* (London: Routledge, 1987), p. 131.
5. See my *In Search of P. D. Ouspensky* (Wheaton, IL; Quest Books, 2003), *Swedenborg: An Introduction to His Life and Ideas* (New York: Tarcher/Penguin 2012), and *Rudolf Steiner: An Introduction to His Life and Work* (New York: Tarcher/Penguin, 2007).
6. 'Waking Sleep: The Hypnagogic State' can be found here: http://www.mindpowernews.com/Hypagogic.htm
7. Gary Lachman, 'Dreaming Ahead', The Quest, Vol. 85 No.12, Winter 1997, pp. 18–23.
8. Jorge Luis Borges, *Labyrinths* (Harmondsworth, UK: Penguin Books, 1976), pp. 87–94.

Chapter One: I Had a Dream Last Night

1. P. D. Ouspensky, *A New Model of the Universe* (New York: Alfred A. Knopf, 1969), p. 254.
2. Gary Valentine, *New York Rocker: My Life in the Blank Generation* (New York: Thunder's Mouth Press, 2006).
3. Gary Lachman, *Beyond the Robot: The Life and Work of Colin Wilson* (New York: Tarcher Perigee, 2016).
4. The song was released by Blondie in April 1978. It was later recorded by Annie Lenox, Tracy Ullman, and the band 10,000 Maniacs. A recording of it by my band, the Know, can be found on *Tomorrow Belongs to You* (2003), a CD of my music.
5. https://en.wikipedia.org/wiki/(I%27m_Always_Touched_by_Your)_Presence,_Dear
6. J. W. Dunne, *An Experiment with Time* (London: Macmillan, 1981), p. 44.
7. Dean Radin, *Real Magic* (New York: Harmony Books, 2018), pp. 87–91.

8. Oddly enough, to add to my collection of odd coincidences, I have spent a good deal of this morning, before getting down to this writing, posting notices about a talk I will give on May 8, 2020, and remarking in the posts that May 8 is White Lotus Day, when theosophists around the world commemorate the death of Madame Blavatsky, who died on that day in 1891. I did not know the date of the Mount Pelée eruption until just now.

9. The problem with this simple but inadequate explanation for why we dream what we do is that we see many things each day, but only a few, or even one, may be in our dreams that night. The question remains why, out of a plurality of elements, this one in particular caught your dream artist's attention.

10. J. B. Priestley, *Man and Time* (London: Aldus Books, 1964), p. 200.

11. https://www.japantimes.co.jp/news/2000/06/04/national/residents-pray-for-victims-of-1991-unzen-eruption/#.XqbhpGhKhPY

12. It still exists: https://www.chadotearoom.com/

13. While going over my dream journals from the early 1990s, I saw that in many dreams from that time I am in London. I am giving tourists directions, asking about exchange rates between US and UK currency, am at pubs and other clearly recognisable British milieus, and am even unsure of when to use '-ise' or '-ize' in words like 'realize' (US usage) and 'recognise' (UK usage). Although I have been an Anglophile since growing up with the Beatles and the 'British Invasion' of the 1960s, this last conundrum has been something I've repeatedly faced since moving to London in 1996 and becoming a full time writer, with copy editors on both sides of the pond correcting my lapses into either British or American English. Although I cannot pin any particular dream down as a clear precognitive one of my living in London, the '-ise' or '-ize' question is not one I had to face until I did move here and started writing fulltime. If this was a presage of my relocating, then the London dreams of the early '90s may have been precognitive pre-echoes of my move there in 1996.

14. For US spelling of 'realize' see preceding note.

15. If I wanted to push the analogies here, I could say that the title of the other piece of music to be played in the dream concert, *Verklärte Nacht*, means in English 'Transfigured Night'. I could suggest that this is a commentary on the dream itself, as it seems, to me at least, to be a very good example of symbolic distortion in precognitive dreams, and hence the night during which I had the dream was one of 'transfiguration'.

16. I even mentioned this in the talk: https://www.youtube.com/watch?v=ekrrwaG_DD4

17. https://twitter.com/GaryLachman/status/1233894458455928832. The article had nothing to do with Wilson or *The Outsider* and was, I later discovered, about a pop singer, Billie Eilish, about whom I am sadly ignorant.

18. https://twitter.com/GaryLachman/status/1244281518496849922.

19. They appear in footnotes in my *Jung the Mystic* (New York: Tarcher/Penguin, 2010).

Chapter Two: The Nightly Sea Journey

1. Anthony Stevens, *Private Myths: Dreams and Dreaming* (Cambridge, MA: Harvard University Press. 1995), p. 8.

2. Literature is not the only art indebted to dreams. In his memoirs, Richard

Wagner records that the opening bars of 'Das Rheingold', the first part of *Der Ring des Nibelungen*, came to him in a dream.

3. Arthur Koestler, *Janus: A Summing Up* (London: Hutchinson, 1978), p. 149.

4. Havelock Ellis, *The World of Dreams* (Boston: Houghton & Mifflin Co., 1922), p. 41.

5. P. D. Ouspensky, *A New Model of the Universe* (New York: Alfred A Knopf, 1969), p. 249.

6. Ellis, p. 51.

7. ibid. p.63.

8. J. B. Priestley, *Rain Upon Godshill* (London: William Heineman Ltd., 1939), p. 298; William James, 'A Suggestion About Mysticism' at https://www.jstor.org/stable/pdf/2011271.pdf.

9. Peter and Elizabeth Fenwick, *The Hidden Door: Understanding and Controlling Dreams* (London: Headline, 1997), p. 3.

10. Koestler, 1978, p. 151.

11. J. B. Priestley, *Man and Time* (London: Aldus Books, 1964), p. 301.

12. Fenwick, p. 3.

13. There is of course no evidence for this, at least of a written kind, but there is reason to believe that our evolutionary ancestors, the Neanderthals, dreamed more than we do and may even have built – if that is the correct word for it – an entire 'dream civilisation', a thesis presented by the psychologist and parapsychologist Stan Gooch in a series of books in the 1970s and 1980s. Gooch contends that the cerebellum, the earlier 'little brain' over which our modern – in evolutionary terms – cerebrum has grown, is the neurological home of the unconscious mind and dreams. Gooch contends that Neanderthals had a much larger cerebellum than Cro-Magnon, our immediate evolutionary ancestor, and that they spent a great deal of time in a state of consciousness similar to what we experience as dreaming. In *A Secret History of Consciousness* I look at Gooch's ideas as related to those of philosophers of the evolution of consciousness such as Rudolf Steiner and Jean Gebser, both of whom suggest that at an earlier stage of human evolution something like Gooch's 'dream consciousness' was dominant. See Stan Gooch *Cities of Dreams* (London: Aulis Books, 1995). An example of a modern 'dream culture' can be found in Kilton Stewart's classic paper on the Senoi of the Malaya Peninsula, collected in Charles T. Tart, ed. *Altered States of Consciousness* (New York: Anchor Books, 1969), pp. 161–171. This collection also includes Frederik Van Eeden's important essay 'A Study of Dreams', in which he coins the term 'lucid dream' (pp. 147–160), and Tart's own study of the 'high dream', a dream in which one experiences the effects of psychedelic drugs (pp. 171–176).

14. A very readable and informative work on dream incubation is Jennifer Dumpert's *Liminal Dreaming* (Berkeley, CA: North Atlantic Books, 2019).

15. Stevens, 1995, p. 15.

16. Ellis, p. 63.

17. Fenwick, p. 3.

18. David Coxhead and Susan Hiller, *Dreams: Visions of the Night* (London: Thames and Hudson, 1990), p. 8.

19. William James 'A Suggestion About Mysticism', at https://www.jstor.org/stable/pdf/2011271.pdf.

20. Stevens, 1995, p. 17.

21. Fyodor Dostoyevsky, *The Brothers Karamazov*, trans. Constance Garnet (New

York: Modern Library, n.d.), p. 538.

22. In Greek the word for 'horn' is similar to that for 'fulfilment'; the word for 'ivory' is similar to that for 'deceive'. So dreams that come through the gate of horn are fulfilled; those that pass through the gate of ivory are deceptive.

23. Fenwick, p. 30. This should not surprise us, as Crick considers our conscious experience, including our sense of having free will, is 'no more than the behaviour of a vast assembly of nerve cells and their associated molecules.' Francis Crick, *The Astonishing Hypothesis* (London: Simon and Schuster, 1994), p. 3.

24. Marie-Louise von Franz, *Dreams* (Boston, MA: Shambhala, 1991), p. 37.

25. http://classics.mit.edu/Aristotle/dreams.html.

26. Ouspensky, p. 261.

27. Anthony Stevens, *Ariadne's Clue* (London: Allen Lane, 1998), p. 16.

28. See Anthony Stevens *Archetypes Revisited* (London: Routledge 2002).

29. An argument against Locke and in favour of Jung is the fact that babies in the womb dream. If dreams are the product of stimuli reaching the mind through the senses, unborn children have yet to have any such experience. And although they do go through cycles of wakefulness and sleep, their periods of wakefulness cannot be altogether that different from those of sleep. So if they dream, what are they dreaming of? Jung would say the archetypes. They are more or less rehearsing for what will greet them when they emerge from the womb.

30. Colin Wilson, *Mysteries* (London: Watkins, 2006), p. 511.

31. Stan Gooch, *Personality and Evolution* (London: Wildwood House, 1971), p. 97.

32. Stevens, 2002, p. 17.

33. In 1954, when he was seventy-nine, Jung said that he had 'carefully analysed about 2000 dreams per annum'. *C. G. Jung: Letters* ed. Gerhard Adler, vol 2. (London: Routledge, 1974), p. 192. But then people came to him with their dreams; Artemidorus had to hunt them down.

34. Freud was dismissive of anything to do with mysticism or the occult, yet he did believe in telepathy and some form of precognition, but kept these ideas mostly to himself. See Gary Lachman 'Was Freud Afraid of the Occult?' in Fortean Times #350, February 2017.

35. Brian Inglis, *The Power of Dreams* (London: Grafton Books, 1987), p. 32. Inglis' book also contains examples of how dreams have informed scholarship, sports, even chess.

36. Mavromatis, p. 192.

37. ibid. p. 186.

38. Von Franz, 1991, takes a Jungian approach to Descartes' dream – too long to quote here – and that of Socrates, mentioned above.

39. In 1930, Pauli posited the existence of the neutrino in order to account for the loss of energy in the decay of atomic nuclei. The Pauli Exclusion Principle states that no two electrons can be simultaneously in the same state or configuration.

40. C. G. Jung, *Dreams* (Princeton, NJ: Princeton University Press, 1974), pp. 111–297.

41. A similar approach is taken by those interested in lucid dreams, in which we can be awake and to some degree 'control' the dream, can act in it with some agency and so confront certain fears and anxieties. This is without doubt an important and useful approach. One concern is that if, as Jung and others suggest, dreams are a natural product of the psyche and often show us parts of ourselves with which we are not well acquainted, can directing a dream to serve our purposes interfere with its autonomy?

42. William Dement, who was involved in the early stages of this research, later founded the Sleep Research Centre at Stanford University.

43. Stevens, 1995, p. 88.

44. The neuroscientists Denis Pare and Rodolfo Llinas noted that the forty hertz oscillations associated with consciousness occur during REM sleep. They concluded that the only difference between our dreaming and waking states is that in waking states 'the closed system that generates oscillatory states' is modulated by incoming stimuli from the outside world. We could say then that what we experience as our waking state, is a kind of dream with a sensory topping. See my *A Secret History of Consciousness* (Great Barrington, MA: Lindisfarne Books, 2003), pp. xix–xxx.

45. Hervy de Saint-Denys, *Dreams and How to Guide Them* ed. Morton Schatzman (London: Duckworth, 1982), p. 52.

46. ibid. p. 89.

47. In fact, one question men suffering from erectile dysfunction are asked is if they wake up with an erection. If so, then the problem is psychological, not physical.

48. Stan Gooch argues that the notion of 'falling asleep', with the suggestion of moving from a height to somewhere lower, is a linguistic expression of the actual 'movement of consciousness' from the cerebral cortex, home of waking, rational thought, to the evolutionary earlier cerebellum, the 'little brain', which the cerebrum obscures and which Gooch believes is the source of dreams. See Chapter 1, note 3 above and Stan Gooch *Total Man* (London: Abacus, 1975), pp. 226–31.

49. Stephen LaBerge and Howard Rheingold, *Exploring the World of Lucid Dreaming* (New York: Ballantine Books, 1991), p. 21.

50. 'Hypnopompic' was coined by the early parapsychologist F. W. H. Myers, to indicate the intermediary state rising up from sleep. If the hypnogogue leads us into sleep, the hypnopomp leads us out of it.

51. Stevens, 1995, p. 91.

52. T. C. Lethbridge, *The Power of the Pendulum* (London: Arkana, 1986), p. 36.

53. Coxhead and Hiller, p. 4.

54. Gary Lachman, *Lost Knowledge of the Imagination* (Edinburgh, UK: Floris Books, 2017), p. 9.

55. Henri Ellenberger, *The Discovery of the Unconscious* (London: Fontana Press, 1994).

56. Georg Christoph Lichtenberg, *The Lichtenberg Reader* ed. Franz Mautner and Henry Hatfield (Boston: Beacon Press, 1959), p. 16.

57. Signe Toksvig, *Emanuel Swedenborg, Scientist and Mystic* (London: Faber and Faber, 1948), pp. 86–7.

58. J. B. Priestley, *Instead of the Trees* (London: William Heineman, LTD, 1977), p. 104.

59. Lachman, 2012, pp. 73–74.

60. Wilson Van Dusen, *Introduction to Emanuel Swedenborg Journal of Dreams* ed. William R. Woofenden, (London and Bryn Athyn, PA: Swedenborg Society and Swedenborg Scientific Association, 1989), p. xxxi.

61. William Beckford, *Vathek and Other Stories* (London: Penguin Books, 1995), pp. 191–223.

62. Friedrich Nietzsche, *Human, All Too Human*, trans. R.J. Hollingdale (Cambridge, UK: Cambridge University Press, 1986), p. 17.

63. Stevens, 1995, p. 31.

64. Colin Wilson, *Introduction to Catharine Crowe The Night-Side of Nature* (Wellingborough, UK: Aquarian Press, 1989), p. ix.

Chapter Three: It's About Time

1. Joan Stambaugh, Introduction to her translation of Martin Heidegger *On Time and Being* (New York: Harper and Row, 1972), p. x.
2. One famous tussle over this question involved the philosophers J. E. McTaggart, who denied time existed, and G. E. Moore, who affirmed that it did. Iris Murdoch quipped that 'McTaggart says that time is unreal, Moore replies that he has just had his breakfast.' Iris Murdoch *Existentialists and Mystics* (London: Chatto & Windus, 1997), p. 299. An interesting and entertaining attempt to deny time's reality can be found in Jorge Luis Borges' 'A New Refutation of Time' in *Labyrinths* (Harmondsworth, UK: Penguin Books, 1976). Borges was influenced by J. W. Dunne and Ouspensky.
3. Stevens, 1995, p. 28.
4. Quoted in J. B. Priestley *Rain Upon Godshill* (London: William Heinemann Ltd 1939), p. 307.
5. Alfred North Whitehead, *Modes of Thought* (New York: The Free Press, 1968), p. 1.
6. Lethbridge, 1986, p. 6.
7. Joan Forman, *The Mask of Time* (London: Corgi Books, 1981), p. 16.
8. Danah Zohar, *Through the Time Barrier* (London: Heineman, 1982), p. 2.
9. Quoted in Dean Radin *The Noetic Universe*, (London: Corgi Books, 2009), p. 107.
10. Quoted in Alister Hardy, Robert Harvie, Arthur Koestler *The Challenge of Chance* (London: Hutchinson, 1973), p. 232.
11. Quoted in Ian Nicholson 'Mutable Time' in *The Book of Time* ed. Colin Wilson (Newton Abbot, UK: Westbridge Books, 1980), p. 162.
12. ibid.
13. Whitehead had his own ideas about relativity, and proposed an alternate version to Einstein's. https://www.openhorizons.org/whitehead-einstein-and-relativity.html
14. Newton and Leibniz were involved in a feud over who 'discovered' calculus first. https://www.thegreatcoursesdaily.com/invented-calculus-newton-leibniz/
15. Quoted in Nicholson, 1980, p. 157.
16. ibid. p. 158.
17. Priestley, 1964, p. 27.
18. Arthur Schopenhauer *Essays and Aphorisms* trans. R.J. Hollingdale (London: Penguin Books, 1988), p. 43.
19. George Steiner, *Martin Heidegger* (Chicago: University of Chicago Press, 1987), p. 110.
20. Søren Kierkegaard, *The Concept of Eternity* trans. Reidar Thomte (Princeton, NJ: Princeton University Press, 1980), p. 85.
21. See Arnold's 'Dover Beach' and Tennyson's 'The Kraken'.
22. Anthony Storr, *Churchill's Black Dog* (London: Collins, 1989), p. 89.
23. Michael Shallis, *On Time* (London: Burnett Books LTD, 1982), p.16.
24. Colin Wilson, *From Atlantis to the Sphinx* (London: Virgin Books, 1996), p. 206.
25. Again, see the work of Stan Gooch on our Neanderthal inheritance.
26. Wilson, 1996, p. 214.
27. Charles Baudelaire, *Paris Spleen*, trans. Louise Varèse (New York: New Directions, 1970), p. 74.
28. Albert Camus, *The Myth of Sisyphus*, trans. Justin O'Brien (London: Penguin Books, 2000), p. 19.

29. Mircea Eliade, *The Sacred and the Profane* (New York: Harcourt, Brace and World, 1959), p. 68.

30. ibid., p. 69.

31. ibid., p. 70.

32. Maurice Nicoll, *Living Time* (London: Watkins Publishing, 1981).

33. Eliade, 1959, p. 71.

34. Wilson, 1996, p. 239.

35. P. D. Ouspensky, *Tertium Organum* (New York: Alfred A. Knopf, 1981), p. 166.

36. C. G. Jung, *Memories, Dreams, Reflections* (London: Fontana, 1989), p. 267.

37. http://nautil.us/issue/35/boundaries/ this-philosopher-helped-ensure-there-was-no-nobel-for-relativity

38. Michael Foley, *Life Lessons from Bergson* (London: Pan Macmillan, 2013), p. 25.

39. Quoted in Wilson, 1996, pp. 240–41.

40. In my book on Rudolf Steiner I briefly mention another occasion when I felt my consciousness was 'cradling' a rose, as if my perception of the rose had shifted from simply seeing it to somehow holding it with my mind. Again, this was a way of slowing down consciousness, which usually is satisfied with taking a 'snapshot' of the world. See *Rudolf Steiner: An Introduction to His Life and Work* (Edinburgh, UK: Floris Books, 2007), p. xxi.

41. I learned of this experiment in Itzhak Bentov's *Stalking the Wild Pendulum* (Rochester, VT: Inner Traditions, 1988).

42. Colin Wilson, *Frankenstein's Castle* (Sevenoaks, UK: Ashgrove Press, 1980), p. 20.

43. Specifically in *The Secret Teachers of the Western World* (New York: Tarcher/Penguin, 2015), and *The Caretakers of the Cosmos* (Edinburgh, UK: Floris Books, 2013).

44. Colin Wilson 'Time in Disarray' in Wilson ed. 1980.

45. Michael Polanyi *The Tacit Dimension* (New York: Anchor Books, 1967), p. 4.

46. Wilson, 1980, p. 303.

47. ibid.

48. Friedrich Nietzsche *Untimely Mediations* (Cambridge, UK: Cambridge University Press, 1983) p. 60. In *The Return of Holy Russia* (Rochester, VT: Inner Traditions, 2020) I examine at length some of the drawbacks to an overly right brain consciousness.

49. Aldous Huxley, *The Doors of Perception* and *Heaven and Hell* (London: Grafton Books, 1987), p. 18.

50. ibid., p. 20.

51. Julian Jaynes, *The Origin of Consciousness in the Breakdown of the Bicameral Mind* (Boston: Houghton Mifflin Company, 1976).

52. This refers to the odd fact that the left brain controls the right side of the body while the right brain controls the left.

Chapter Four: Looking Ahead

1. Colin Wilson, *The Psychic Detectives* (New York: Berkeley Books, 1987), p. 26.

2. Colin Wilson, *Mysteries* (London: Watkins Publishing, 2006), p. 167.

3. Louisa Rhine, *Hidden Channels of the Mind* (London: Victor Gollancz, 1962), pp. 27–28.

4. F. W. H. Myers, *Human Personality and Its Survival of Bodily Death* (New York:

Dover Books, 2005), p. 102.
5.   Stan Gooch, *The Paranormal* (London: Fontana 1979), p. 43.
6.   Zohar, 1982, p. 2.
7.   Hardy, Harvie, Koestler, 1973.
8.   https://www.scientificamerican.com/article/according-to-current-phys/
9.   Algernon Blackwood, *Shocks* (London: Grayson and Grayson, 1935), pp. 245–50.
10. https://www.sciencealert.com/scientists-propose-a-mirror-universe-where-time-moves-backwards
11. H.G. Wells, *The Shape of Things to Come* (London: Hutchinson, 1933), p. 15.
12. I am indebted to Colin Wilson and Stan Gooch who voice these arguments in several of their books.
13. Radin, 2018, p. 96. I can't help but mention that Colin Wilson voiced something similar half a century ago in *The Occult* when he said that scientists who dismiss the paranormal as absurd 'are closing their eyes to evidence that would convince them if it concerned the mating habits of albino rats or the behaviour of alpha particles.' Wilson, 2015, p. xli.
14. Steven Weinberg, *The First Three Minutes* (New York: Basic Books, 1993), p. 154.
15. Henri Bergson, *The Two Sources of Morality and Religion* (New York: Doubleday & Co., 1935), p. 317.
16. https://koestlerunit.wordpress.com/
17. Arthur Koestler, *The Roots of Coincidence* (New York: Random House, 1972), p. 13.
18. ibid. p 50.
19. The Committee for the Scientific Investigation of Claims of the Paranormal. Some of its members were Carl Sagan, the Amazing Randi (a stage magician), Christopher Evans, Martin Gardner, and John Wheeler.
20. Gary Lachman, *The Caretakers of the Cosmos* (Edinburgh, UK: Floris Books, 2013), pp.190–92. Wheeler's theory involves the famous 'twin slit photon' experiment and the 'collapse of the wave function' and is too involved to relate here. Steiner made his remark in *Goethe's Worldview* at https://wn.rsarchive.org/Books/GA006/English/APC1928/GA006_index.html
21. Rhine, 1962, p. 7.
22. Quoted in Koestler, 1972, p. 14.
23. Radin, 2018, p. 135.
24. Gooch, 1979, pp. 176, 195.
25. Radin, pp. 129, 136–40.
26. Gary Lachman, *Dark Star Rising: Magick and Power in the Age of Trump* (New York: Tarcher Perigee, 2018), p. 27.
27. Eric Wargo, *Time Loops* (San Antonio, TX: Anomalist Books, 2018), p. 70.
28. ibid. p. 70.
29. ibid. p. 71.
30. ibid. p. 217.
31. ibid. p. 42.
32. I should mention that precognition is accepted in an even more stringent environment than science – the business world. There is what is known as the 'precog economy', in which financiers employ 'seers' to advise on investments. If high rollers and the military are interested in this sort of thing, it must be real. See the *Observer* September 29, 2019 https://www.theguardian.com/global/2019/sep/29/

psychic-future-what-next-for-the-precognition-economy?CMP=share_btn_tw.

33. Colin Wilson, *Beyond the Occult* (London: Watkins Publishing, 2008), p. 211.
34. Wargo, 2018, p. 31.
35. https://www.youtube.com/watch?v=gQP13EQ0dJo.
36. Steve Taylor, *Making Time* (Thriplow, UK: 2008), p. 158.
37. https://www.newyorker.com/magazine/2019/03/04/
     the-psychiatrist-who-believed-people-could-tell-the-future.
38. https://twitter.com/GaryLachman/status/1244283559818539010
39. Dunne, 1981, p. 42.
40. J. W. Dunne, *Intrusions* (London: Faber and Faber, 1955), p. 100.
41. Lethbridge, 1986, p. 8, p. 40.
42. J. W. Dunne, *Nothing Dies* (London: Faber and Faber, 1940), pp. 24–30.
43. I am reminded here of the 'incompleteness theorem' of the Austrian logician
     Kurt Gödel. Simply put, it states that any formal axiomatic system has inherent
     limitations which require something outside the system to resolve. In a sense we
     can say that Gödel's incompleteness theorem puts paid to any idea of a 'theory of
     everything'.
44. P. D. Ouspensky, *A Further Record* (London: Arkana, 1986), p. 1; Ouspensky,
     1969, p. 294.
45. Ouspensky, 1969, pp. 370–80.
46. Anthony Peake, *Time and the Rose Garden* (Winchester, UK: 2018), p. 14.
47. Priestley, 1977, pp. 105–06.
48. Priestley, 1939, p. 309.
49. ibid. pp. 290–91.
50. Priestley, 1964, p. 188.
51. ibid. p. 258.
52. ibid. p. 46.
53. ibid. p. 221.
54. Ouspensky, 1968, p. 418.
55. Ouspensky's ideas about recurrence form the plot of the film *Groundhog Day*.
56. Michael Shallis recounts an experience in which he tried to alter a future while in
     the act of becoming aware of it. He tells of a powerful experience of *déjà vu* while
     teaching a physics lesson. At one point the *déjà vu* feeling came over him and
     he knew that he was going to suggest he retrieve a book from his office to show
     some examples. He decided to resist the suggestion. But no sooner did he make
     this resolution than he heard himself say that he had better get the book. (Shallis,
     pp. 159–60). What I can say here is that not everyone alters the outcome of a
     precognitive dream in the way that Priestley's camper did. In some cases it isn't
     necessary. In some, the attempt to do so ensured the outcome would happen; this
     is the gist of several fatalistic tales about the impossibility of escaping destiny. But
     some do. Also, all choices are not equally important. Trying not to retrieve a book
     is not the same as making sure your child doesn't drown. And perhaps his student
     really needed the examples and so he would be acting against his best interests if
     he didn't get the book. From my own experience I can say that the *déjà vu* feeling
     is not the same as the 'I dreamed this' recognition. I have experienced both many
     times and can only say that the flavour of *déjà vu* is not the same as precognition.
     In one I say, 'I feel this has happened before.' In the other I say, 'I dreamed
     this' and know that I have. The *déjà vu* 'tingle' lasts for a few moments. The
     recognition that 'I dreamed this' is a sudden shock. I can also attest to deciding
     to allow something I had dreamed about to happen. In Chapter One I relate a

precognitive dream in which my sons' mother cycled past while I sat outside at a café. As soon as she arrived, the dream of sitting at a picnic table came to me. I thought 'I can ask if she would like a coffee or not.' I felt clearly that it was up to me. I did not, as Shallis did, try to prevent the foreseen event from happening, but I did feel that I could decide either way.

57. Priestley, 1977, p. 93.
58. For a good introduction to Lethbridge's work see Colin Wilson's *Mysteries*, whose first four chapters are, in effect, a short book about precisely that.
59. Wilson, 2006, p. 55.
60. Lethbridge, 1984, pp. 115–16, 122–23.
61. ibid. p. 19.
62. ibid. p. 6.
63. ibid. p. 48.

Chapter Five: What a Coincidence

1. Taylor, 2008, p. 154.
2. But see Robert Hardie, 'Probability and Serendipity', in Hardy, Harvie, Koestler, 1973.
3. Arthur Koestler, *The Case of the Midwife Toad* (New York: Random House, 1971), p. 135.
4. https://twitter.com/GaryLachman/status/1263455003588669444
5. *The Hands of Orlac* (1924), an Austrian silent horror film directed by Robert Wiene, is a recurring motif in Lowry's novel.
6. https://encyclopedia.ushmm.org/content/en/article/chiune-sempo-sugihara
7. See Perle Epstein, *The Private Labyrinth of Malcolm Lowry* (New York: Holt, Rhinehart, Winston, 1969), for a look at the occult influences informing Lowry's masterpiece. See also Gary Lachman, *The Dedalus Book of the Occult* (Sawtry, UK: Dedalus Books, 2003), pp. 251–59.
8. Hardy, Harvie, Koestler, 1973, p. 186.
9. Colin Wilson with Damon Wilson, *The Encyclopaedia of Unsolved Mysteries* (Chicago, Il: Contemporary Books, 1988), p. 262.
10. Hardy, Harvie, Koestler, 1973, p. 187.
11. ibid. p. 178.
12. Koestler, 1978, p. 268.
13. Koestler's 'ink-fish effect' is related to William James' 'law' regarding psychic phenomena: that there will always be enough evidence to confirm believers but never enough to convince sceptics.
14. Hardy, Harvie, Koestler, 1973, p. 215.
15. In books like *Life and Habit* (1878) and *Unconscious Memory* (1880), Butler argued against Darwin, who he saw as 'banishing mind from the universe.'
16. https://www.sheknows.com/entertainment/slideshow/9629/celebrity-death-rule-of-threes/
17. Koestler, 1972, p. 85.
18. ibid.
19. ibid. p. 86.
20. ibid.
21. Koestler, 1978, pp. 266–67.

22. Gary Lachman, *The Return of Holy Russia* (Rochester, VT: Inner Traditions, 2020), pp.342–49.

23. ibid. p. 87.

24. Gary Lachman, *Jung the Mystic* (New York: Tarcher/Penguin, 2010).

25. C. G. Jung foreword to Richard Wilhelm trans. *The I Ching or Book of Changes* (New York: Bollingen Foundation, 1977) p. xxiv.

26. Shallis, 1982, p. 147.

27. That the number of hexagrams in the *I Ching* is the same as the number of the possible combinations of the code signs in the RNA and DNA molecules has struck some as a remarkable synchronicity. See Michael Hayes, *The Infinite Harmony* (London: Weidenfeld & Nicolson, 1994).

28. Wilhelm, 1977, pp. 20–21.

29. Jung, ibid. p. xxv.

30. R.F.C. Hull, ed. *C. G. Jung Speaking* (Princeton, NJ: Princeton University Press, 1987), p. 183.

31. Marie-Louise von Franz, *Psyche and Matter* (Boston: Shambhala, 1992), p. 26.

32. Priestley, 1964, p. 163.

33. F. David Peat, *Synchronicity: The Bridge Between Matter and Mind* (New York: Bantam Books, 1988).

34. C. G. Jung, 'Synchronicity: An Acausal Connecting Principle' in Jung and Pauli, *The Interpretation and Nature of the Psyche* (New York: Pantheon Books, 1955), p. 123.

35. Koestler, 1972, p. 98.

36. Jung, 1955, p. 124.

37. ibid.

Chapter Six: A Telescope into the Past

1. C. E. M. Joad, *Guide to Modern Thought* (London: Pan Books, 1943), pp. 208–10.

2. ibid. p. 210.

3. Phillipe Jullian, *Prince of Aesthetes: Count Robert de Montesquiou, 1855–1921*, (New York: Viking Press, 1965).

4. Wilson, 2006, p. 359.

5. Joan Forman, *The Mask of Time* (London: Corgi Books, 1981), p. 78.

6. ibid. p. 24.

7. http://friends-of-fotheringhay-church.co.uk/history/

8. Joad, 1943, p. 210.

9. T. C. Lethbridge, *The Essential T. C. Lethbridge* (London: Routledge & Kegan Paul, 1980), p. 4.

10. ibid. p. 10.

11. http://www.hows.org.uk/personal/hillfigs/lost/cambri.htm

12. Lethbridge, 1980, p. 11.

13. Colin Wilson, *The Psychic Detectives* (New York: Berkeley Books, 1987). I am indebted to Wilson's book for much in this account.

14. ibid. p. 31.

15. Wilson gets this phrase from David Foster, a cybernetics expert whom he writes

about in *The Occult*. Priestley knew Foster too and mentions him in *Instead of the Trees*, p. 38.

16. Lachman, 2012.
17. Colin Wilson, 2015, pp. 75–77.
18. Quoted in ibid. p.77.
19. Priestley, 1939, p.325, 1937 p. 275.
20. In *Lost Knowledge of the Imagination* I recount Goethe's similar use of imagination. Lachman, 2017, pp. 60–61.
21. Colin Wilson, *The Philosopher's Stone* (Los Angeles: Jeremy P. Tarcher, 1989), pp. 114–15.
22. Lachman, 2016, p. 109.
23. Forman, 1981, p. 53.
24. Wilson, 1980, pp. 290, 313. Wilson has an interesting argument for why time travel in Wells' sense is a misconception. He argues that there is no 'time' in the sense of Newton's abstract equable flow. 'Time' is the name we give to process, what Leibniz meant when he said that 'instants, considered without the things, are nothing at all.' If this is the case, then it is meaningless to speak of 'process travel'. He compares what we experience as time to what we see looking out the window on a train journey. We see the landscape 'flowing' past. It doesn't really; what we see as a 'flow' is really an effect of the train moving through the landscape. If we call this apparent flow 'zyme' and spoke of 'travelling in zyme' we can see that this is an example of what Whitehead called 'misplaced concreteness', when language leads us to imagine as an independent reality what is really the product of something else.
25. Koestler, 1978, p. 48.
26. Priestley, 1964, p. 294.
27. Ouspensky, 1969, p. 290.
28. James 'A Suggestion about Mysticism' at https://www.jstor.org/stable/pdf/2011271.pdf
29. Quoted in Wilson, 1987, p. 258.
30. Priestley, 1939, p. 278.
31. Wilson, 1980, p. 312.

Bibliography

Baudelaire, Charles, *Paris Spleen*, trans. Louise Varèse, New York: New Directions, 1970.

Beckford, William, *Vathek and Other Stories*, London: Penguin Books, 1995.

Bentov, Itzhak, *Stalking the Wild Pendulum*, Rochester, VT: Inner Traditions, 1988.

Bergson, Henri, *The Two Sources of Morality and Religion*, New York: Doubleday & Co., 1935.

Blackwood, Algernon, *Shocks*, London: Grayson and Grayson, 1935.

Borges, Jorge Louis, *Labyrinths*, Harmondsworth, UK; Penguin Books, 1976.

Camus, Albert, *The Myth of Sisyphus*, London: Penguin Books, 2000.

Coxhead, David and Hiller, Susan, *Dreams: Visions of the Night*, London: Thames and Hudson, 1990.

Crowe, Catherine, *The Night Side of Nature*, Wellingborough, UK: Aquarian Press, 1989.

Dostoyevsky, Fyodor, *The Brothers Karamazov*, trans. Constance Garnet, New York: Modern Library, n.d.

Dumpert, Jennifer, *Liminal Dreaming*, Berkeley, CA: North Atlantic Books, 2019.

Dunne, J. W., *An Experiment With Time*, London: Macmillan, 1981.

—, *Intrusions*, London: Faber and Faber, 1955.

—, *Nothing Dies*, London: Faber and Faber, 1940.

Eliade, Mircea. *The Sacred and the Profane*, New York: Harcourt, Brace and World: 1959.

Ellenberger, Henri, *The Discovery of the Unconscious*, London: Fontana Press, 1994.

Ellis, Havelock, *The World of Dreams*, Boston: Houghton and Mifflin, 1922.

Fenwick, Peter and Elizabeth, *The Hidden Door: Understanding and Controlling Dreams*, London: Headline, 1997.

Foley, Michael, *Life Lessons From Bergson*, London: Pan Macmillan, 2013.

Forman, Joan, *The Mask of Time*, London: Corgi Books, 1981.

Gooch, Stan, *Cities of Dreams*, London: Aulis Books, 1995.

—, *The Paranormal*, London: Fontana, 1979.

—, *Personality and Evolution*, London: Wildwood, House, 1971.

—, *Total Man*, London: Abacus, 1975.

Heidegger, Martin, *On Time and Being*, trans. Joan Stambaugh. New York: Harper and Row, 1982.

Hull, R. F. C., *C. G. Jung Speaking*, Princeton, NJ: Princeton University Press, 1987.

Huxley, Aldous, *The Doors of Perception and Heaven and Hell*, London: Grafton, Books, 1987.

Inglis, Brian, *The Power of Dreams*, London: Grafton Books, 1987.

Jaynes, Julian, *The Origin of Consciousness in the Breakdown of the Bicameral Mind*, Boston: Houghton & Mifflin Co., 1976.

Joad, C. E. M., *Guide to Modern Thought*, London: Pan Books, 1943.

Jullian, Phillipe, *Robert de Montesquiou, King of Aesthetes*, New York: Viking Press, 1965.

Jung, C. G., *Dreams*, Princeton, NJ: Princeton University Press, 1974.

—, *C. G. Jung: Letters*, ed. Gerhard Adler, London: Routledge, 1974.

—, *Memories, Dreams, Reflections*, London: Fontana, 1989.

Jung, C. G., with Pauli, Wolfgang, *The Interpretation and Nature of the Psyche*, New York: Pantheon Books, 1955.

Kierkegaard, Søren, *The Concept of Anxiety*, Princeton, NJ: Princeton University Press, 1980.

Koestler, Arthur, *The Case of the Midwife Toad*, New York: Random House, 1971.

—, *Janus A Summing Up*, London: Hutchinson, 1978.

—, *The Roots of Coincidence*, New York: Random House, 1972.

Koestler, Arthur, with Hardy, Alistair and Harvey, *The Challenge of Chance*, New York: Random House, 1973.

Laberge, S. and Rheingold, H., *Exploring the World of Lucid Dreaming*, New York: Ballantine Books, 1991.

Lachman, Gary, *Beyond the Robot: The Life and Work of Colin Wilson*, New York: Tarcher Perigee, 2016.

—, *The Caretakers of the Cosmos*, Edinburgh: Floris Books, 2013.

—, *Dark Star Rising: Magick and Power in the Age of Trump*, New York: Tarcher Perigee, 2018.

—, *In Search of P. D. Ouspensky*, Wheaton, Il: Quest Books, 2006.

—, *Jung the Mystic*, New York: Tarcher/Penguin, 2010.

—, *Lost Knowledge of the Imagination*, Edinburgh: Floris Books, 2017.

—, *The Return of Holy Russia*, Rochester, VT: Inner Traditions, 2020.

—, *Rudolf Steiner*, New York: Tarcher/Penguin, 2007.

—, *A Secret History of Consciousness*, Great Barrington, MA: Lindisfarne Books, 2003.

—, *The Secret Teachers of the Western World*, New York: Tarcher/Penguin, 2015.

—, *Swedenborg*, New York: Tarcher/Penguin, 2012.

(As Gary Valentine) *New York Rocker*, New York: Thunder's Mouth Press, 2006.

Lethbridge, T. C., *The Essential T. C. Lethbridge*, London: Routledge and Kegan Paul, 1980.

—, *The Power of the Pendulum*, London: Arkana, 1984.

Lichtenberg, George Christophe, *The Lichtenberg Reader*, ed. Franz Mautner and Henry Hatfield, Boston, MA: Beacon Press, 1959.

Mavromatis, Andreas, *Hypnagogia*, London: Routledge, 1987.

Myers, F. W. H., *Personality and Its Survival of Bodily Death*, New York: Dover Books, 2005.

Nicoll, Maurice, *Living Time*, London: Watkins Publishing, 1981.

Nietzsche, Friedrich, *Human, All Too Human*, trans. R.J. Hollingdale, Cambridge, UK: Cambridge University Press.

—, *Untimely Meditations*, trans. R.J. Hollingdale, Cambridge, UK: Cambridge University Press.

Ouspensky, P. D., *A Further Record*, London: Arkana, 1986.

—, *A New Model of the Universe*, New York: Alfred A. Knopf, 1969.

—, *Strange Life of Ivan Osokin*, New York: Hermitage House, 1955.

—, *Tertium Organum*, New York: Alfred A. Knopf, 1981.

Peake, Anthony, *Time and the Rose Garden*, Winchester, UK: O Books, 2018.

Peat, F. David, *Synchronicity: The Bridge Between Matter and Mind*, New York: Bantam Books, 1988.

Polanyi, Michael, *The Tacit Dimension*, New York: Anchor Books, 1967.

Priestley, J. B., *Instead of the Trees*, London, William Heineman, 1977.

—, *Man and Time*, London: Aldus Books, 1964.

—, *Over the Long High Wall*, London: William Heineman, 1972.

—, *Rain Upon Godshill*, London: William Heineman, 1939.

Radin, Dean, *The Noetic Universe*, London: Corgi Books, 2001.

—, *Real Magic*, New York: Harmony, Books, 2018.

Rhine, Louisa, *Hidden Channels of the Mind*, London: Victor Gollancz, 1962.

Schopenhauer, Arthur, *Essays and Aphorisms*, trans. R. J. Hollingdale, London: Penguin Books, 1988.

Shallis, Michael, *On Time*, London: Burnett Books, 1982.

Steiner, George, *Heidegger*, Chicago: University of Chicago Press, 1987.

Stevens, Anthony, *Ariadne's Clue*, London: Allen Lane, 1998.

—, *Private Myth: Dreams and Dreaming*, Cambridge, MA: Harvard University Press.

Storr, Anthony, *Churchill's Black Dog*, London: Collins, 1989.

Swedenborg, Emanuel, *Journal of Dreams*, ed. William R. Woofenden. London and Bryn Athyn, PA: Swedenborg Society and Swedenborg Scientific Association, 1989.

Tart, Charles, ed., *Altered States of Consciousness*, New York: Anchor Books, 1969.

Taylor, Steve, *Making Time*, Thriplow, UK: Icon Books, 2008.

Toksvig, Signe, *Emanuel Swedenborg*, London: Faber and Faber, 1948.

Von Franz, Marie-Louise, *Dreams*, Boston, MA: Shambhala, 1991.

—, *Psyche and Matter*, Boston, MA: Shambhala, 1992.

Wargo, Eric, *Time Loops*, San Antonio, TX: Anomalist Books, 2018.

Wells, H. G., *The Shape of Things to Come*, London: Hutchinson, 1933.

Whitehead, Alfred North, *Modes of Thought*, New York: Free Press, 1968.

Wilhelm, Richard trans., *The I Ching or Book of Changes*, New York: Bollingen Foundation, 1977.

Wilson, Colin, *Beyond the Occult*, London: Watkins Books, 2008

—, *The Book of Time*, ed., Newton Abbot, UK: Westbridge Books, 1980.

—, *The Encyclopaedia of Unsolved Mysteries*, with Wilson, Damon. Chicago: Contemporary Books, 1988.

—, *Frankenstein's Castle*, Sevenoaks, UK: Ashgrove Press, 1980.

—, *From Atlantis to the Sphinx*, London: Virgin Books, 1996.

—, *Mysteries*, London: Watkins Books, 2006.

—, *The Occult*, London: Watkin Books, 2015.

—, *The Philosopher's Stone*, Los Angeles: Jeremy P. Tarcher, 1989.

—, *The Psychic Detectives*, New York: Berkeley Books, 1987.

Zohar, Danah, *Through the Time Barrier*, London: William Heineman, 1982.

Index

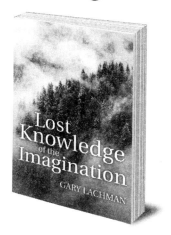

The Caretakers of the Cosmos
Living Responsibly in an Unfinished World

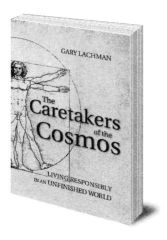

"A worthy and important book."
The Journal of the Unitarian Society for Psychical Studies

"A cracking book on a cracking subject by a cracking author."
Magonia Review of Books

Why are we here? Human beings have asked themselves this question for centuries. Modern science largely argues that human beings are chance products of a purposeless universe, but other traditions believe humanity has an essential role and responsibility in creation.

Gary Lachman brings together many strands of esoteric, spiritual and philosophical thought to form a counter-argument to twenty-first century nihilism, addressing some of the most urgent questions facing humanity.

Also available as an eBook

florisbooks.co.uk

The Quest For Hermes Trismegistus
From Ancient Egypt to the Modern World

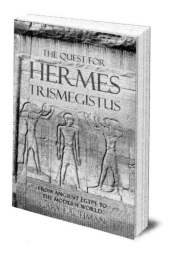

"An excellent addition to any history collection."
Midwest Book Review

"Lachman is an easy to read author yet has a near
encyclopaedic knowledge of esotericism."
Living Traditions Magazine

Considered by some a contemporary of Moses and a forerunner
of Christ, the almost mythical figure of Hermes Trismegistus
was thought to have walked with gods and be the source of
the divine wisdom granted to man at the dawn of time.

Gary Lachman brings to life the mysterious character of this
great spiritual guide, exposing the many theories and stories
surrounding him, and revitalizing his teachings for the
modern world.

Also available as an eBook

florisbooks.co.uk

Rudolf Steiner
An Introduction to His Life and Work

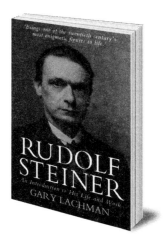

"[Gary Lachman] has rendered a great service
to Steiner and his movement."
Scientific and Medical Network Review

"Brings one of the twentieth century's
most enigmatic figures to life."
Booklist

Rudolf Steiner – educator, architect, artist, philosopher and
agriculturalist – ranks amongst the most creative and prolific
figures of the early twentieth century. Yet he remains a mystery
to most people.

Gary Lachman tells Steiner's story lucidly and with great
insight in the first truly popular biography, written by a
sympathetic but critical outsider. He presents Steiner's key
ideas in a readable, accessible way, tracing his beginning as
a young intellectual to the founding of his own metaphysical
teaching, called anthroposophy.

florisbooks.co.uk

Floris
Books

For news on all our **latest books,**
and to receive **exclusive discounts,**
join our mailing list at:

florisbooks.co.uk

Plus subscribers get a FREE book
with every online order!

We will never pass your details to anyone else.